Chemistry in Introductory Geology

Chemistry in Introductory Geology

FOURTH EDITION

By W. D. Keller

Professor of Geology
UNIVERSITY OF MISSOURI

Lucas
Brothers
Publishers

909 LOWRY
COLUMBIA, MISSOURI 65201

First edition, 1957
Second edition, 1963
Third edition, 1966

CHEMISTRY IN INTRODUCTORY GEOLOGY
Fourth Revised edition

Library of Congress Catalog Card Number 71-90870

Standard Book Number 87543-053-8

Printed in the United States of America

TABLE OF CONTENTS

FOREWORD

Many of the most interesting and important processes of geology are basically chemical reactions between geologic materials, and therefore they are most clearly understood when explained in the terms of chemistry. In order to prepare students in introductory geology for such explanations, a chapter on chemistry is usually included in the textbooks of introductory geology. Unfortunately for the students, and especially for those who have had no previous instruction in chemistry, the first application of chemistry encountered in geology is one of the most difficult of all. For that reason confusion, and often an erroneous impression that geological chemistry is difficult to understand, may result in the minds of the students, whereas if the introduction of chemistry had begun with ideas familiar to them, the application of chemistry to geology would have been enjoyable and rewarding.

The purpose of this booklet is to provide in language easy to read, a simplified but sound approach to the concepts of chemistry used most commonly in introductory geology, and to illustrate the concepts with practical geologic examples. It is written primarily for the geology student who has had no previous course in chemistry. It may be useful also to those who wish to have brought together more closely than is done in geology textbooks the several applications of chemistry in introductory geology, and to those who wish to review simple geological chemistry.

The material included in this booklet represents a compromise between the maximum amount of technical chemistry likely to be encountered in an introductory geology course and that beyond which the presentation may not go without defeating its purpose of being easy to understand.

The writer has benefited in the preparation of this booklet by suggestions from his colleagues, especially Daniel A. Bradley, at the University of Missouri. Mr. Herman Ponder has drafted most of the illustrations.

Second Edition

During preparation of the second edition of this booklet, suggestions for its improvement made by colleagues and users of the first edition were gratefully utilized. Parts on clay mineral structures were deleted, and a summary of the rock cycle was added. Much of the last half of the booklet was rewritten. I am grateful for the reading of the manuscript by Alden Carpenter and Maynard Slaughter.

W.D. Keller
Columbia, Mo.
June, 1963

Fourth Edition

Again I am grateful for comments made by users of this booklet. Particularly at the suggestion of Professor Joseph F. Schreiber, Jr., certain parts of the first edition, after being deleted from the second editon, are returned to this one. Additional new material on the Bowen reaction series, and on weathering by plants has been included.

W.D. Keller
Columbia, Mo.
July, 1969

The Materials of the Earth in Terms of Elements, Atoms, Ions, and Compounds

Geology and chemistry have long been closely related sciences. Early inorganic chemistry began as an outgrowth of mineralogy, and geologic materials have always been the source of most of the raw materials of chemistry. Conversely, geology has received much from chemistry because the reactions by which rocks and minerals originate and change follow the laws formulated by chemists to describe reactions carried out in the laboratory. Thus the two sciences supplement each other indispensably.

The close relationships between geology and chemistry may be illustrated with geologic examples by recalling that the earth's crust (the part of the earth we can see) is composed of rocks and minerals which are, in reality, chemical compounds, some of which are very common and some rare. The earth is, therefore, truly a huge natural storeroom of chemical substances, analogous to a storeroom in a chemistry laboratory. The geologist searches for and quarries from his storeroom the desired minerals of metals and nonmetals. Carrying the analogy further, in various parts of the large earth laboratory erupting volcanoes simulate melting in huge crucibles, the deposition of ore minerals occurs by chemical precipitation, and falling rain represents large scale distillation and condensation.

The changes which rocks, minerals, and ores undergo in the earth are similar to what the chemist carries on in the laboratory under the name "chemical reactions." A chemical reaction proceeds always in a certain direction which can be predicted provided the conditions of the reacting system are adequately known. Because there is no fundamental difference if the reaction is in a beaker or in the wide ocean, the laws of chemistry may be used to predict and describe geologic reactions. In the succeeding pages, we will endeavor to interrelate geology and chemistry in many ways, and to explain why and how, in terms of demonstrable experiments, that geologic materials and changes, many of which can not be manipulated by us because of their huge size or remote location, occur as they do in the earth.

Elements

The earth is composed of rocks and minerals which are composed fundamentally of chemical elements. A chemical element may be defined, for our purpose, as a substance which can not be changed to something simpler by ordinary chemical processes.* It is essential for geologists to have not only a definition of an element but also a feeling for, and a workable understanding of, what an element is. We will begin to familiarize ourselves about chemical elements with a metal that is composed of not one, but several elements.

Everyone is familiar with common brass, like that used in certain door knobs, low-priced jewelry, and so forth. Suppose a piece of such brass is sent to a chemist for an analysis. In the analytical procedure he will separate the components of the brass metal and then identify them by appropriate tests. He will report the presence of copper, zinc, and tin in the originally single metal brass.

Next suppose the analyst takes the individual fractions of copper, zinc, and tin which he has separated from the brass and sends them to a second chemist for confirmation. Assuming that both chemists have been competent, the second chemist will report that the copper is copper, zinc is zinc, and the tin is tin. This confirmation will be repeated as often as competent analyses are made.

* One element may be changed into another by nuclear reaction but this process is considered to be not ordinary for our purpose. A more rigorous definition of an element is a substance with one atomic number. (See page 7).

It is now apparent that the copper, zinc, and tin are different from the brass from which they were separated. Copper, zinc, and tin have not been subdivided into any other substances, whereas brass was broken down and subdivided. The copper, zinc, and tin maintained their uniqueness with complete persistence despite all of the work which chemists did on them. They show, thereby, a fundamental or elemental property of not being changed and are, therefore, called elements.

Elements are not solely metals, as in the examples given; nonmetal elements, such as sulfur and iodine, and gases, such as oxygen and neon, are equally important. There are 93 naturally occurring chemical elements which have been found in the earth's crust, and after the advent of the atomic pile and nuclear reactor, twelve more (to 1969) elements have been produced by nuclear reaction and have been identified. Plutonium was first identified as a synthesized element but later was found in small quantity in natural uranium ore minerals. Our interest as geologists is directed mainly to the naturally occurring elements, and because less than half of them are met with in everyday life, a list of only 37 of the common elements is compiled in Table 1. Complete lists may be found in textbooks of chemistry.

The first eight elements (oxygen to magnesium, inclusive) listed in Table 1 are the most abundant elements in the earth's crust, and make up about 98.5 per cent of it by weight. Fourteen of the elements, which have been marked with asterisks because they will be referred to most often in this booklet, comprise more than 99.5 per cent, by weight, of the earth's crust. It is desirable to learn the chemical symbols (O, Si, Al, etc.) along with the names of the elements because the symbols are regularly used (as a sort of chemical shorthand) to designate compounds and minerals and to describe reactions. Capital and lower case letters should be used as given in the table. Fe for iron is taken from ferrum, Latin, as are Na from natrium, K from kalium, Pb from plumbum, Ag from argentum, Au from

TABLE 1

The Commoner Chemical Elements in the Earth's Crust

Element	Symbol	Weight percentage of earth's crust	Relative weight of elements (atoms)	Number of atoms percentage in crust	Radius in angstrom units. See footnote	Volume percentage of earth's crust
*Oxygen	O	46.60	16	62.55	1.40	91.97
*Silicon	Si	27.72	28.1	21.22	0.39	0.80
*Aluminum	Al	8.13	27	6.47	0.57	0.77
*Iron	Fe	5.00	55.8	1.92	0.82	0.68
*Calcium	Ca	3.63	40	1.94	1.06	1.48
*Sodium	Na	2.83	23	2.64	0.98	1.60
*Potassium	K	2.59	39.1	1.42	1.33	2.14
*Magnesium	Mg	2.09	24.3	1.84	0.78	0.56
Titanium	Ti	.44	47.9			
*Hydrogen	H	.14	1			
*Phosphorus	P	.118	31			
Manganese	Mn	.100	54.9			
*Sulfur	S	.052	32.1		.77	
*Carbon	C	.032	12		.77	
*Chlorine	Cl	.031	35.5		1.81	
Fluorine	F	.030	19			
Strontium	Sr	.030	87.6			
Barium	Ba	.025	137.4			
Chromium	Cr	.020	52			
Vanadium	V	.015	51			
Zinc	Zn	.013	65.9			
Nickel	Ni	.008	58.7			
Copper	Cu	.007	63.5			
Tungsten	W	.0065	183.9			
Lithium	Li	.0065	6.9			
*Nitrogen	N	.0046	14			
Tin	Sn	.004	118.7			
Cobalt	Co	.0023	58.9			
Lead	Pb	.0016	207.2			
Molybdenum	Mo	.0015	96			
Uranium	U	.0004	238			
Bromine	Br	.00016	80			
Mercury	Hg	5×10^{-5} †	200.1			
Iodine	I	3×10^{-5}	126.9			
Silver	Ag	1×10^{-5}	107.9			
Gold	Au	5×10^{-7}	197.2			
Platinum	Pt	5×10^{-7}	195.2			

Data in this table are taken, by permission, from Principles of Geochemistry by Brian Mason, John Wiley and Sons, and from Handbook of Chemistry and Physics, Chemical Rubber Publishing Co.

One Angstrom unit is equivalent to .0000001 mm. There are 25.4 mm. per inch. The radii are given for the elements as they ordinarily occur in rocks and minerals.

(*) The fourteen elements most commonly referred to and written in symbol form in this book are marked with asterisks.

(†) This type of notation is a short method of expressing a fraction; 5×10^{-5} may be written also as 5/100,000, or as 0.00005.

aurum, and Hg from hydrargyrum, and O, Si, Al, and others, come from their names in the English language.

A few of the names of elements may be new to the student but most of them are sufficiently common and familiar that we already have an intuitive feel for them. Oxygen is the gas which makes up approximately 20 per cent of the air. In chemical combination with other elements it is the most abundant element in the earth's solid crust. Silicon, perhaps a less familiar name than oxygen, is the element second most abundant in the earth's crust; it is a substance whose luster is somewhat like that of silver. It has some properties common to metals and some of the nonmetals. Aluminum and iron (elements) are metals common to the experience of almost everyone. Magnesium is a lightweight metal, somewhat similar to aluminum. Calcium, sodium, and potassium are "silvery" metals which react readily with water, even with that which is moisture in the air.

Hydrogen, the lightest of the elements, is a highly flammable gas. Phosphorus is a soft, pale yellow solid which may ignite spontaneously in air; therefore, it is regularly stored under water. Phosphorus compounds are present in bones and teeth; phosphorus is an essential element in the nutrition of plants and animals. Sulfur is a yellow, glistening solid. The fumes of burning sulfur, as well as the rotten-egg odor of hydrogen sulfide, are known to many persons. Carbon is the major constituent of black coal and the only element in pure diamond. Chlorine is a yellowish green gas which is so highly poisonous that small amounts of it will kill bacteria in water. Nitrogen comprises about 78 per cent of ordinary air. Compounds of nitrogen are essential in plant nutrients.

Many of the other elements in the table are familiar because they are metals, or non-metals, in common use. Some of them occur in small amounts, as shown toward the lower part of the third column, and for this reason may command a high economic value to man.

Atoms

In order to compare the weights and various other properties of elements, and to relate quantitatively the combination of one element with another, it is necessary to use a common unit of measure for the elements. The most convenient unit for this purpose is the atom. An atom will be thought of as the smallest portion of an element that possesses the properties of that element. Atoms may be visualized, therefore, as the tiniest packet of an element which may exist individually or in combination with other atoms. Chemists have determined the relative weights of these individual atoms, which are presented to the nearest decimal in column 4 of Table 1. An atom of carbon has a standard reference weight of 12 (unspecified as to the unit) and on the same basis or unit, an atom of hydrogen has a weight of 1, an atom of iron weighs 55.8, and an atom of common uranium 238.

Common practice is to express these weights in grams, whereupon 12 grams of carbon is the gram-atom of carbon. It likewise follows that the same number of atoms in any element yields the gram-atom of that element. This number, called Avogadro's number is quite large: 6.02×10^{23}. That is, 6.02×10^{23} atoms of carbon weigh 12 grams, 6.02×10^{23} atoms of oxygen weigh 16 grams, etc., throughout the table of chemical elements.

The relative number of atoms in the earth's crust, expressed in terms of 100 atoms (i.e., per cent) is shown in column 5 of Table 1. The number per cent of atoms in the crust is necessarily different from the weight per cent of elements (column 3) because the weight of the atoms of each element is different from that of the others.

In column 6 are given the sizes of the elements as ions (atoms carrying excess electrical charges, to be discussed later), the way in which elements commonly occur in rocks and minerals. Because atoms and ions are exceedingly tiny, their radii are regularly expressed in Angstrom units (see footnote to Table 1). Although these units are much smaller than those familiarly used in common experience with geologic measurements, the relative sizes between ions are easily understood. One of the largest of the common ions (elements) is oxygen, which likewise has been listed as the most abundant element in the earth's crust. For these reasons, oxygen makes up over 91 per cent of the volume, or 46.6 per cent of the weight, of the earth's crust, see column 7. The rocky crust of the earth is actually an oxygen (ion) crust.

INTERNAL STRUCTURE OF ATOMS. Although means are not available to magnify atoms so that their components are visible to man, definite conclusions have been drawn about their internal structures from indirect observations.

Each atom is believed to contain a central assembly of matter called a nucleus. Every nucleus contains one or more particles each of which carries a positive electric charge; these particles are called protons. The positive charges in the nucleus of an atom characterize that atom as being a particular kind of element, and the number of positive charges in the nucleus is taken as the atomic number of the element. This is the basis of the definition of an element as being a substance with one atomic number (see page 2). Atomic numbers range from one, that of hydrogen; to 92 in uranium. Other particles, called neutrons, similar in mass to the protons but lacking the positive charge, are present in atomic nuclei (except that of hydrogen). Mesons, of which there are several varieties, function as binding materials in the nucleus, and other particles, such as positrons and neutrinos, not immediately important geologically, have been discovered. All of the particles in the nucleus have been grouped under the name nucleons. The protons and neutrons hold greatest significance (among the particles in the nucleus) in explaining the behavior of elements in geologic processes. Because protons and neutrons are relatively heavy, almost all of the mass of an atom resides in the nucleus.

The remaining portion of the atoms consists of a group of electrically negative charges called electrons which are visualized as occurring in shells or orbits about the nucleus. (See Figure 1.) The electrons in the outer orbits of elements play an important role in chemical reactions because they are transferred between elements during reaction and combination. Before describing further the orbital distribution of electrons, however, it may be interesting to inquire into the relative sizes and masses of protons and electrons.

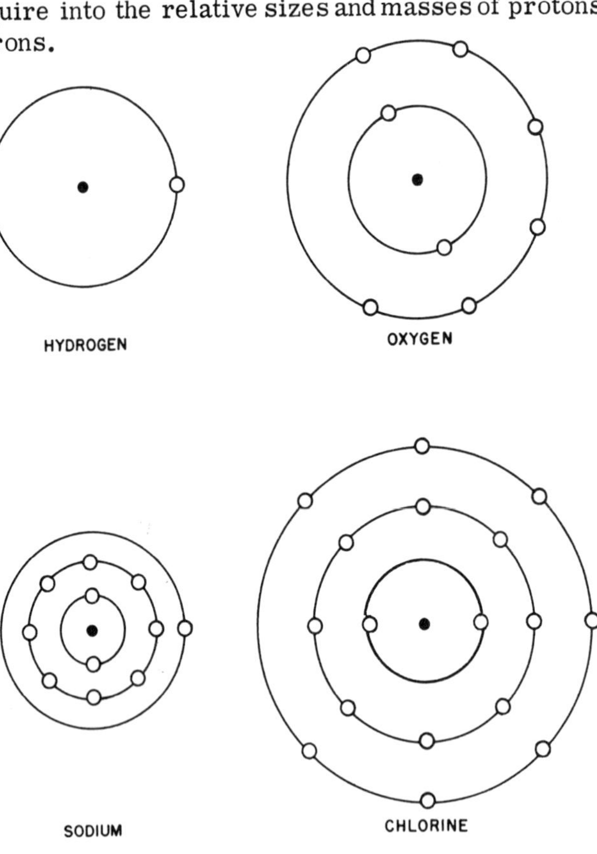

Figure 1. Diagrams of the inferred structure of atoms showing the nucleus at center in black, and electrons as small open circles distributed in orbits, or shells, about the nucleus. Electrons probably are clouds of negative charge concentrated about positions indicated by open circles.

A hydrogen atom, the simplest of atoms, is composed of one proton and one electron, and has a mass of approximately 1.67×10^{-24} grams.* Thus the weight of a hydrogen atom is an almost incomprehensibly small amount, but the mass of an electron is only 1/1837 of that of the hydrogen atom! The electron moves very rapidly (about 2.18×10^{8} cm. per second) about the nucleus and describes thereby a spherical path approximately 2 Angstrom units in diameter.† The nucleus occupies only a very small fraction of the spherical space of the atom. To illustrate, if the sphere described by the moving electron of the hydrogen atom is magnified so that its diameter approximates that of a basketball, the nucleus will still be less than .001 inch in diameter.

An atom contains as many electrons as there are positive charges (protons) in the nucleus. The electrons are distributed about the nucleus in what has been previously described as orbits or shells. Although the distribution of electrons about a nucleus is not necessarily spherical, probably the electron groups sweep out spherical shapes in space. These electron shells occur at various distances from the nucleus, and represent levels of energy by which the electrons are held by the nucleus. The intensity of energy diminishes outward from the nucleus. From two to 32 electrons occupy each major electron shell, following a highly systematic order. An oxygen atom, for example possesses 8 electrons in two shells which contain 2 and 6 electrons respectively, in outward direction, see Figure 1, whereas a gold atom possesses 79 electrons distributed outward from the nucleus in major shells as follows: 2, 8, 18, 32, 18, 1.

The details of electron distribution within each element

* One gram is equivalent to .035 oz. The number 1.67×10^{-24} means the fraction $1.67/10^{24}$, or 1.67 divided by 10 multiplied by itself 23 times, that is, the number 1 followed by 24 zeros.

† The expression 2.18×10^{8} means 2.18 multiplied by the number which is obtained when 10 is multiplied by itself seven times, namely, 2.18 x 100,000,000, or 218,000,000. Angstrom unit has been defined in footnote 3 below Table 1.

are not immediately important in beginning geology and therefore will not be gone into. They are important in understanding certain intricacies of reaction from the chemist's viewpoint, but for us we will note, at this time, only that, although inner shells may contain 32 electrons, the outer shell of electrons contains not more than eight electrons. Indeed, apart from helium in which two electrons are in the outer shell, the most stable configuration of the electrons in an outer shell of atoms is eight. Therefore, the least reactive (most stable) elements are those whose atoms contain eight electrons in their outer shell, and it follows that very important effects accompany the arrangements of eight electrons in the outer shells of the atoms.

As the number of protons (positive charges) in the nucleus increases, and likewise the number of electrons (negative charges) in the entire atom increases, the number of electrons in the outer shell ranges from one to eight in the atoms of the different elements. For example, sodium atoms have only one electron in their outer shells, but chlorine has seven. The atoms of these elements can achieve a more stable state by combining with other atoms in such a way that the electrons in their outer shells will be transferred one to the other, or be mutually shared, with the result that each atom then controls, or shares control of, eight electrons in its outer shell. These combinations will be described subsequently in the discussion of chemical compounds.

Ions

Returning to the examples of sodium and chlorine mentioned above, it is possible for sodium to lose the one electron in its outer shell, whereupon the next innermost shell of electrons, containing eight of them, becomes the outer shell. Certain stability is thereby conferred upon the new configuration. When a sodium atom thus loses one negatively charged electron the sodium residue is left with an excess positive charge, and is written "Na^+" which is a sodium ion. The superscript "+" indicates that the sodium carries one unit of excess positive charge. Sodium occurs as ions in the rocks of the earth's crust and in the ocean.

Chlorine, which as an electrically neutral atom contains seven electrons in its outer shell, exemplifies a slightly different type of atom than sodium. The chlorine atom tends to pick up an extra electron to add to its seven outermost ones, whereupon it will then control eight electrons in its outermost shell. Upon acquiring the extra electron, that chlorine carries one excess unit of negative electrical charge, and therefore becomes Cl^-, a chloride ion. The superscript "-" refers to the negative charge.

Hence, we may think of ions as atoms which carry electrical charges. Ions with negative charges carry excess electrons, whereas ions with positive charges are deficient in electrons with respect to their nuclear charges. Ions are exceedingly important in geological chemistry because they (ions), rather than neutral atoms, participate in most rock and mineral changes and in geological materials and reactions. Indeed, geologic chemistry has been described as the chemistry of migrating ions in the earth.

The ionic charge assumed by each element is characteristic for that particular element, but different elements give rise to ions of different charges in accord with their different electron configurations. The calcium atom contains two electrons in its outer shell; aluminum, three; and silicon, four electrons. They, and others, assume the corresponding number of positive charges to become ions, as shown in the following group: H^+, K^+, Ca^{++}, Mg^{++}, Al^{+++}, Si^{++++}. Iron may lose either two or three electrons, depending upon the surrounding chemical environment, and become, respectively, Fe^{++} or Fe^{+++} (page 41).

The oxygen atom contains six electrons in its outer shell, and adds two to become the oxygen ion, O^{--}. Oxygen in combination with other elements also forms common groups of ions, such as, CO_3^{--} (called "carbonate"), HCO_3^- (bicarbonate), SO_4^{--} (sulfate). If this array of ions appears formidable to the student at first contact, he should not become dismayed, or necessarily memorize all of them at once, because they will become familiar to him as they are used later in connection with substances which are familiar to him.

It is noteworthy that the properties of an ion may be radically different from those of its corresponding atom. Attention has already been directed to oxygen ions as a major constituent of hard, heavy rocks, although oxygen gas is tenuous. Hence, oxygen ions in combination with other ions give rise to compounds entirely different in appearance from mechanical mixtures of their neutral atoms mingled together. Likewise ions of sodium and chlorine in water constitute a solution of edible ordinary table salt, whereas chlorine gas alone is poisonous, and atoms of sodium constitute a metal. Other examples of even more spectacular and exciting differences between ions and atoms of the same element could be cited.

Chemical Compounds

If Na and Cl are combined chemically to form solid NaCl, they do so by becoming ions and combining as ionic crystalline rock salt. Na loses an electron to become Na^+, and Cl picks up the "lost" electron to become Cl^-. Hence,

Figure 2. Crystals of NaCl, the mineral halite from rock salt. Note the cubic form, which arises from the arrangement of ions in the crystal.

NaCl is bonded physically and chemically by the transfer of electrons. This is called an ionic bond. Both Na and Cl ions in NaCl hold eight electrons in their outer shells, a "preferred", more stable state.

NaCl is the formula for ordinary table salt. Large crystals of NaCl, as shown in Figure 2, contain countless numbers of Na^+ and Cl^- arranged very precisely in a definite geometrical pattern. The drawing and model of NaCl Figure 3 (a) and (b), illustrate the alternate positions of Na^+ and Cl^- in the crystal of salt, i.e., the compound NaCl. The external cubical shape and right-angle cleavage of table salt crystals is obviously an inheritance of the internal arrangement of the constituent Na and Cl ions.

The distance from the center of a Na ion to the center of a Cl ion is approximately the sum of their radii: 1.81A (Cl^-) plus .98A (Na^+) (from Table 1), equals 2.79A. Each Na^+ is adjacent to six nearest neighbors of Cl^-, and each Cl^- has six Na^+ as nearest neighbors in a crystal of NaCl. Therefore no single Na^+ "belongs" to any single Cl^-, but the electrical influence of each ion is three dimensional with respect to all others nearby.

Carbon (C) atoms contain four electrons in their outer shells. In order to control eight electrons, each atom must gain four electrons. Carbon has the interesting property whereby a given atom of carbon can share its four electrons mutually with neighbor carbon atoms. The shared electrons provide for a total of 8 in the outer shells of the adjacent nearest neighbors (carbons), a type of bonding known to the chemist as covalent bonding. This type of bonding is exceedingly strong; it provides for the hardness of diamond–crystalline carbon–the hardest substance occurring naturally.

Oxygen ions, O^{--}, combine with Si^{++++} ions in the ratio 2 oxygen: 1 silicon, in part by the NaCl mechanism and in part by the diamond mechanism of bonding. It is interesting that certain ions may substitute for others having nearly the same size in the formation of minerals and

(a)

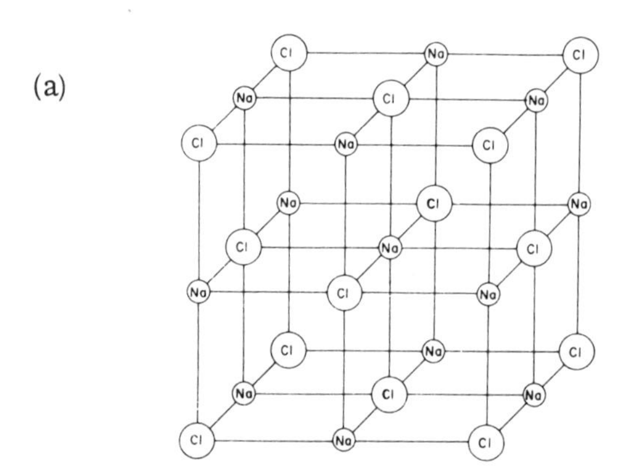

(b)

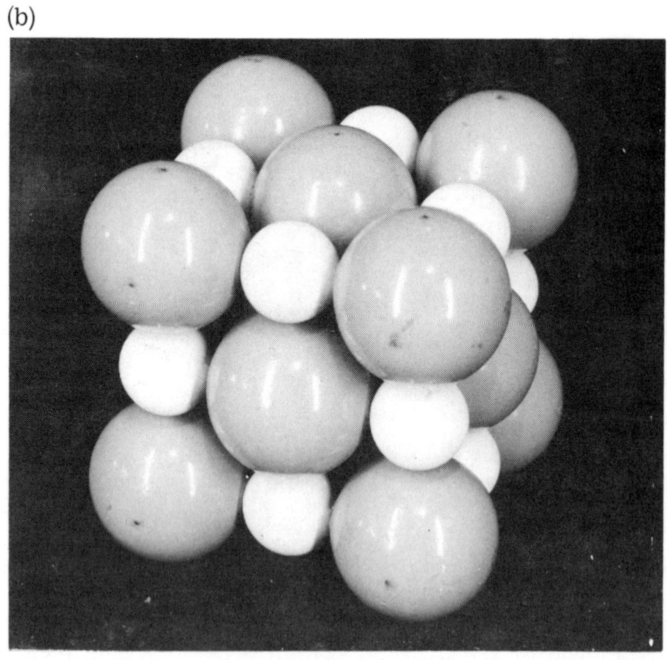

Figure 3. 3(a) is an "exploded" diagram of the arrangement of Na and Cl ions in a crystal of halite. 3(b) is a scale model of small Na ions and large Cl ions in NaCl.

other chemical compounds. For example, ions of Fe and Al, Fe and Mg, Ca and Na, Ag and Au, and many others may proxy for each other, but the total electrical charge in the mineral must be kept in balance (page 32).

Chemical compounds may exist also as liquids, for example, water, which is a compound of two gases, hydrogen and oxygen. The ions and atoms in liquids occur typically in random arrangement; they are not highly organized as was shown in the model of a crystal of rock salt. Likewise, the ions of substances dissolved in liquids, like salt dissolved in water, are dispersed essentially at random.

When chemical elements combine with one another, the resulting substance is known as a chemical compound. Minerals fall into this category. The compound (or mineral) is designated by writing together the chemical symbols for the combining elements with no spaces between them, for example, NaCl. Full significance is to be attached to the quantitative relationship of the two elements as expressed in NaCl. This formula means that equal numbers (a ratio of 1:1) of Na and Cl atoms or ions are combined. The formula for common water is H_2O, which signifies that the ratio of H atoms to O atoms is 2:1. A subscript on an element designates the number of times an atom of that element is taken to make up the simplest unit of the compound or mineral. If no subscript is shown, a value of one (1) is implied. If HO or OH are written (a 1:1 ratio of O:H), they refer to something different from water which is H_2O.

Formulas for some minerals are so lengthy or complex that the effort to learn them in exact detail is greater than the benefits to be had from such effort. For those minerals, formulas in this booklet are commonly written qualitatively, that is, to show which elements are present but not how much of each one, by adding a small (lower case) "n" as a subscript after the formula, which is to mean that unspecified ratios of the elements are taken in the formula. For example, the formula $CaMgFeAlSiO_n$ (for the mineral augite of the pyroxene group) with subscript "n" means not that the ratio of elements is one throughout but rather that the ratio of each to the others is unspecified. The formula is correct elementally or qualitatively, but is non-committal quantitatively. We are now prepared to observe the relationships of ions of elements in geologic materials and reactions.

Summary

An element is a substance that cannot be changed to something simpler by ordinary chemical processes; it has one atomic number. Atoms are characterized by a positively charged central nucleus, and electrons distributed at certain shells (energy levels) about it. Ions are atoms which carry unbalanced electrical charges. Ions combine to form liquid and crystalline compounds.

References for Auxiliary Reading

General Chemistry by H. H. Sisler, C.A. Vanderwerf, and A.W. Davidson, MacMillan Company, New York.

General Chemistry by L. Pauling, W.H. Freeman and Company, San Francisco.

Data of Geochemistry, F.W. Clarke, U.S. Geological Survey Bulletin, No. 770, Superintendent of Documents, Washington, D.C.

Chemical Processes in Geology

Important processes of geology that operate primarily as a result of chemical reactions, or in accord with chemical laws, include crystallization of a magma, weathering of rocks, deposition of sedimentary rocks from solution, deposition of certain ore deposits, metamorphism, and others. Within the discussion of these processes, other topics such as the nature of acids and alkalis, solubilities, oxidation, hydrolysis, etc., will be elaborated using geologic examples.

It will be recalled that geologic processes always operate or occur spontaneously. The most fundamental significance of their spontaneous action is the fact that the products of the processes or reactions possess less energy (typically called "free energy") than did the original participants or reactants. For example, rocks roll downhill and water flows downhill for the basic reason that by doing so they lose energy in the process; they move from a position of higher mechanical potential energy to a position of lower mechanical potential energy. Likewise, a magma cools and solidifies spontaneously because it loses heat to surroundings; in effect it moves "downhill thermally." Many fresh, lustrous rocks and minerals tarnish and weather, and coal burns, these processes being chemical reactions in which "chemical potential energy moves downhill." Indeed, the test for whether any process will occur spontaneously is whether the energy of the participants is diminished during the process.

Geologists are concerned with earth materials (whose compositions are described chemically) and their response to energy (mechanical, thermal, and chemical) which is impressed upon and flows from them.

The Magma and Its Products

Composition and Nature of a Magma

A magma is a hot solution of the ions (see Table 2) which are found in igneous rocks, plus additional water (steam) and other volatile substances. Solid rock and mineral particles may also occur in a magma.

Table 2
Chemical Composition of An Average Igneous Rock

	(a) Weight per cent		(b) Number of ions per cent
SiO_2	59.12	O	60.5
Al_2O_3	15.34	Si	20.6
Fe_2O_3	3.08	Al	6.3
FeO	3.80	Fe'''	.8
MgO	3.49	Fe''	1.1
CaO	5.08	Mg	1.8
Na_2O	3.84	Ca	1.9
K_2O	3.13	Na	2.6
H_2O	1.15	K	1.4
TiO_2	1.05	H	2.6
P_2O_5	.30	Ti	.3
Others	.62	P	.1
	100.00		100.00

The analysis in column (a) was taken from U.S. Geological Survey Professional Paper No. 127.

The chemical composition of magmas in nature is inferred from the chemical analyses of erupting lava and ancient igneous rocks and from synthetic "magmas" prepared in the laboratory. The composition of a computed average igneous rock is given in Table 2, where it is expressed in two ways: as oxides (a custom long followed), and as individual ions, which is probably the more direct way.

The use of oxides (compounds of elements with oxygen and written SiO_2, Al_2O_3, Fe_2O_3, etc.) in describing the chemical composition of rocks, as in column (a) of Table 2, does not mean, necessarily, that those individual oxide compounds existed in that form in the rock. Actually they are not present as minerals having the composition of Al_2O_3 or CaO, etc; instead, the use of oxides to record chemical compositions of rocks originated as an outgrowth of analytical procedures long ago, and its use has continued through custom. Because this form of reporting compositions is widely prevalent it has been reproduced here.

Chemical compositions are also expressed in terms of ions, column (b), which represents the same rock as column (a). This form seems to be more descriptive of the way we visualize a rock or magma to be constituted. A magma is easily visualized as consisting of a body or framework of many large oxygen ions (perhaps as spheres somewhat distorted in shape) in random arrangement which are bonded together by smaller ions (also distorted spheres) of Si, Al, Fe, Ca, K, etc. which occur interstitially between the oxygens. Because the magma is very hot, the ions are in rapid oscillating motion and therefore are in random (not systematic) arrangement. The high mobility of the ions accounts for the magma being a liquid. The magma is a liquid solution of the ions, which means that the ions are intimately and essentially homogeneously mixed. Homogeneity, as a single phase, is an attribute of solutions (solutions are discussed in greater detail on page 42) and the magma is no exception to this property.

The magma may be compared crudely with a boiling hot pan of fudge candy. This is also an homogeneous solution of ingredients from which certain grains will crystallize out upon cooling. The liquid fudge mixture is not a simple melt like that which is obtained when pure sugar is heated; the fudge mixture is a solution of sugar, water, chocolate, etc., which exists as a liquid at a temperature below the melting point of some of its components. When the fudge is later cooled, materials crystallize out of it over a temperature interval as they become relatively insoluble.

The concept that a magma is a solution but not a pure melt is so important that space will be taken to cite another example common to our experience. It is recalled that solid ice and solid salt (being solid, both exist below their melting temperatures) may be stirred together and become liquid at a temperature below the freezing temperature of water. The point here is that a solution may exist as a liquid at temperatures below the melting or liquid temperature of any of the single (pure) ingredients in the solution.

In the same manner, a magma is a hot solution which is a liquid below the melting temperatures of the individual minerals which will crystallize from the magma as it later cools. A concise summary description of a magma, for our purposes, is that a magma is a liquid solution of mineral-forming ions in random arrangement, which possesses a relatively large amount of both thermal and chemical energy. When the energy of the magma is decreased it tends to crystallize.

Cooling and Solidification of the Magma

The rate at which a magma cools, or loses energy, influences the physical chemical reactions which take place in the magma during the cooling interval. We will examine first the effects of very rapid, and later slow, cooling and solidification.

RAPID COOLING AND SOLIDIFICATION OF THE MAGMA. If the magma is chilled very rapidly, as in a thin lava flow, much of the thermal (heat) energy of its ions is lost very rapidly. This rapid loss of heat results in the ions being frozen or fixed in the same random arrangement as that in which they occurred in the magma. The liquid magma has therefore been chilled to a glass, which is a form of matter also described as an undercooled liquid. The glassy rock may be thought of, figuratively, as one stationary picture frame withdrawn from a movie sequence of the mobile liquid magma. The vigorously oscillating ions in the liquid have been relatively fixed (relatively short vibrations) in the "solid" glass. The structure of artificially prepared glasses has been interpreted by x-ray diffraction studies of the glasses and these interpretations lead to the descriptions given above of the structure in the igneous rock glass (such as obsidian) and the magmatic liquid of which it is a counterpart.

SLOW SOLIDIFICATION OF THE MAGMA. The description of the slow solidification of a magma begins with a hot liquid magma possessing relatively large amounts of thermal and chemical energy. The ions are in rapid motion (high kinetic energy) which results in high fluidity. As the magma cools, after emplacement, it loses thermal energy to its surroundings. The loss of thermal energy is in effect a decrease in the vigor of motion of the ions, and consequently a decrease in liquidity of the magma.

During this interval while the motion of the ions is slowly decreasing, a very interesting corollary process is operating simultaneously, namely, a likewise slow, but highly systematic, geometric arranging and grouping of the ions of one kind to another. A highly organized geometric arrangement of ions is an attribute of a crystal (recall that of rock salt, page 14) and therefore when slow solidification of a magma occurs, the ions succeed in moving about sufficiently so as to arrange themselves into crystals. Crystals possess less energy than does either a liquid or glass having the same composition. Therefore, there is (1) a spontaneous tendency of a magma to crystallize, and (2) energy is given off by the formation of crystals. In other words, there is a "downhill movement" of energy in the crystallization process, which explains why crystallization occurs when a magma solidifies slowly.

As the temperature of the magma falls, different minerals crystallize from the magma at different temperatures, which are in reality different levels of energy, thereby setting up a sequence of interrelated mineral types, temperatures of crystallization and reaction, and energies of crystallization. The common rock-forming silicate (silicate means it contains O, Si, and other ions) mineral which crystallizes first, that is, with least temperature decline, from a cooling magma is olivine $(Mg, Fe)_2SiO_4$. As olivine and succeeding metal silicates crystallize, the metal ions are withdrawn from the liquid, thereby enriching relatively the remaining liquid in SiO_2. It follows that the last mineral to crystallize, that is, at the lowest temperature, is quartz, SiO_2.

A mineral which has crystallized at a higher temperature may later react with the liquid magma remaining at a lower temperature, tending to form a new mineral containing more silica which is stable and at equilibrium at the lower temperature. This reaction will be discussed further as the Bowen reaction series on page 33. The importance of energy factors tending to establish equilibrium is thus clearly demonstrated. The sequence of minerals arranged in order of crystallization from magma, and their departure from a high energy level in the magma, is shown diagramatically in Figure 4. (See next page)

A range and sequence of bonding energy in the silicate minerals arises mainly from the relative number of bonds between oxygen ions and silicon ions in them because the Si-O bonds are the strongest in silicate minerals. Bonds between O and Fe, Mg, Ca, and other metals, are weaker than Si-O bonds. The way by which the relative number of Si-O bonds in silicate structures can vary will become evident as the details of the structures of the several silicate groups are discussed.

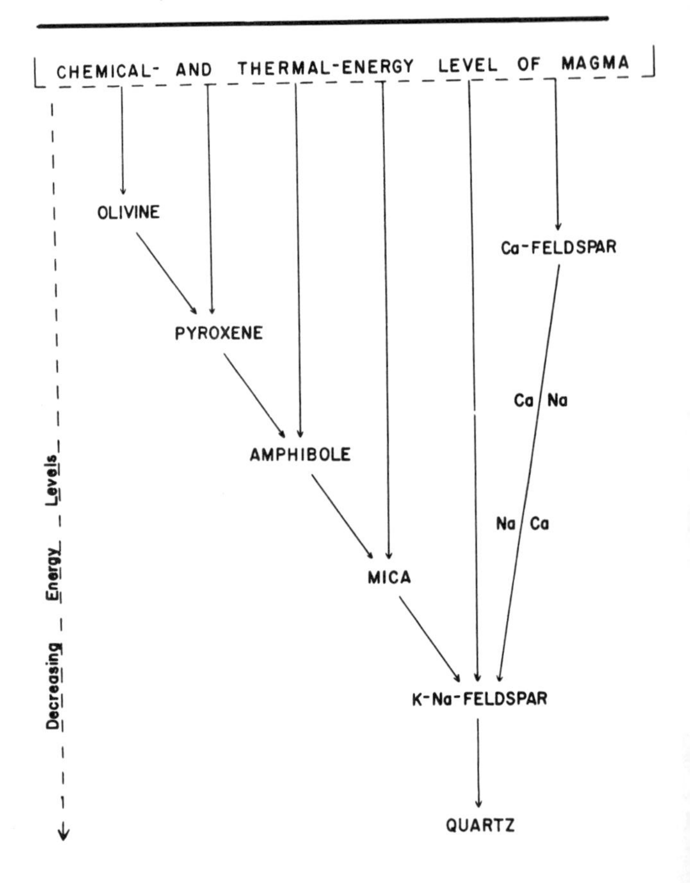

Figure 4. Diagram showing the crystallization of minerals from magma arranged in descending order of energy levels with respect to the original magma.

We begin with the chemical composition of the representative original magma, analysis (b) in Table 2, which shows that O is present in largest number, followed by Si. It has also been pointed out previously that each Si ion carries 4+ charges (page 11) which are readily shared by one of the two minus (-) charges of each of 4 oxygen ions arranged in even distribution about the Si. Figure 5 shows similar tetrahedral models in 2 orientations illustrating this grouping of 4 oxygens about one silicon.

Figure 5. Two models, in different orientation, of silica tetrahedra. The wooden balls represent oxygen. A small sphere representing a silicon ion is located, but hidden by the oxygen balls, in the interstice between the oxygen balls. This and other pictures of silicate structures taken from "Principles of Chemical Weathering", W.D. Keller, Lucas Brothers Publishers, Columbia, Missouri.

The geometry of this 4-cornered, tetrahedral, arrangement, Figure 5, is essentially perfect because 4 oxygen ions placed in tetrahedral position leave in the void between them adequate space for a Si ion to fit snuggly.

This assembly is called a silica tetrahedron, SiO_4^{----}, which is the fundamental building block of the silicate minrals. (See footnote next page) The near-perfect spatial acking in the silica tetrahedron coincides with efficient electrical bonding in it, and these highly compatible relationships account for the tight structure, superior hardness, notable strength, and high luster and melting temperature that characterize most silicate minerals.*

Olivine. Let us resume consideration of the crystallization process of the magma. Olivine crystallizes when silica tetrahedra are formed, and twice as many Mg or Fe ions as the number of silica tetrahedra are available for linking them together. More figuratively, say out of every 100,000 ions in the magma, perhaps 1000 of the oxygens will assemble tetrahedrally about 250 silicon ions to form 250 silica tetrahedra. These in turn are linked to one another by 500 ions of magnesium and iron, forming olivine, Figure 6(a). Each Mg, or Fe'', ion carries a charge of 2^+, these being readily shared by the (-) minus charge which otherwise is unsatisfied on each oxygen of each SiO_4^{----} tetrahedron. If two Mg^{++} ions (Fe^{++} may substitute for Mg^{++}) are available for each SiO_4^{----}, all of the excess electrical charges on each are compensated, and a continuous network composed of SiO_4 tetrahedra bridged between corners by Mg and Fe ions is assembled, see Figure 6 (b).

* One Si ion fits into the interstice between four oxygens arranged in a tetrahedral group because it possesses a compatible size to do so. It can be demonstrated on a purely spatial geometrical basis, entirely apart from ionic, electrical, or compositional properties, that spheres of certain sizes will fit into the voids left between groups of larger spheres which are arranged into specific systematic packings. For example, if the ratio of the radius of the small, interstitial sphere to the radii of the larger enclosing spheres lies between 0.22 and 0.41 to 1, the small sphere will fit between four larger spheres arranged as a tetrahedron. Illustrating with Si and O, the radius of Si was given as 0.39 and that of O as 1.4 (Table 1); the ratio of Si:O is therefore 0.39:1.40, or 0.28:1, which falls between 0.22 and 0.41 to 1. Si should fit on the inside of tetrahedral oxygen, and it does fit as predicted. Silicon is said to have a coordination number of 4 with respect to oxygen because 4 oxygens can be grouped about it.

If the ratio of radii of the small to large spheres falls between 0.41 and 0.73 to 1, sufficient room is available for the small sphere when six of the larger ones are grouped about it in an octahedral pattern. The coordination number of the small sphere is therefore six. An octahedral pattern is one that exhibits eight sides, or a shape indicative of eight sides. The ratio of the radius of Al(0.57) to that of O is about 0.42:1, a value that is in the lower part of the range in octahedral coordination. True to prediction, Al is coordinated octahedrally with O in minerals that crystallize at temperatures near the earth's surface, but at high temperatures, where the ions have relatively high kinetic energy, Al can also be coordinated tetrahedrally, like Si is.

Between ratios 0.73 to 1 of the smaller sphere to 1 of the others, a cubical coordination of eight is geometrically compatible.

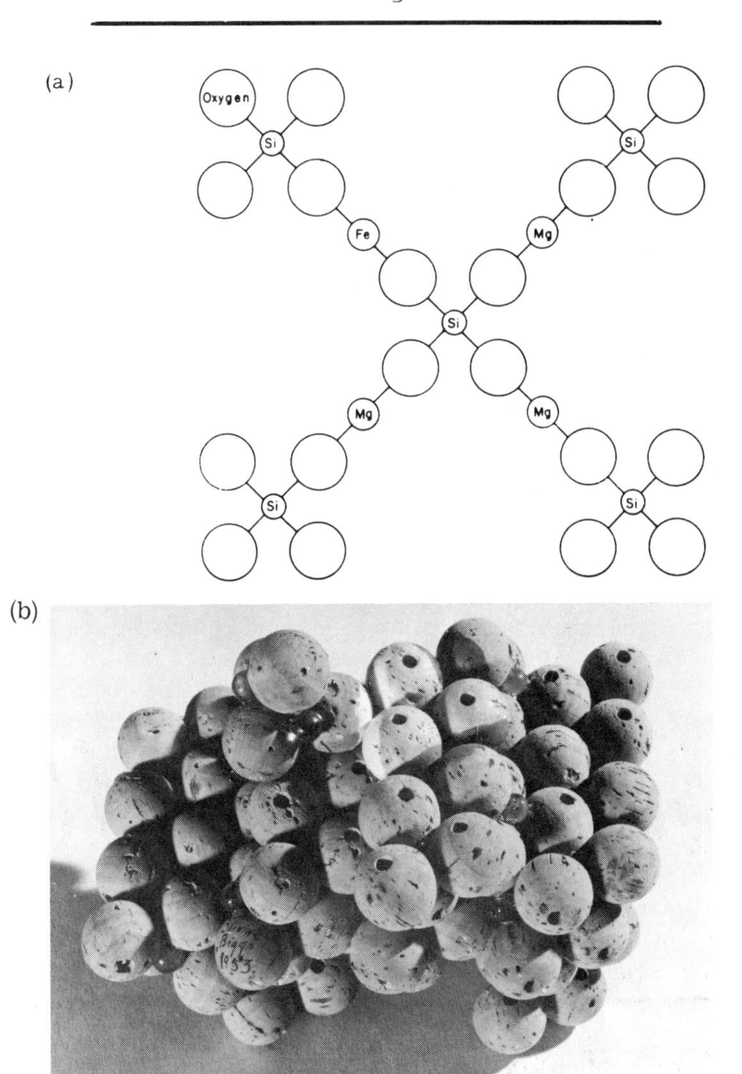

Figure 6. 6(a) illustrates diagrammatically the linking of silica tetrahedra by ions of magnesium and iron in olivine. 6(b) is a three-dimensional model of olivine, the oxygens being represented by cork balls and the magnesium and iron by beads. Silicon is hidden within the tetrahedra. (Model assembled by Dr. D.L. Biggs).

Thus, crystals of olivine that are large represent a correspondingly large number of SiO_4 tetrahedra linked with Mg and Fe. When magma solidifies slowly, the long time interval during which liquid is transformed into solid, (1) permits the ions to arrange themselves into SiO_4, and Mg, Fe bridges, and, (2) allows many, many of these crystal units to form and align themselves systematically thereby growing a large crystal (of olivine). In this way, grains large enough to be seen by the unaided eye, as in the rocks called gabbros, are crystallized from magmas.

In the preceding discussion it was assumed that the concentration of Mg and Fe ions in the magma was high enough to form olivine at the appropriate temperature-energy level. This is a valid assumption for the more simafic (a hybrid word from Si, Mg, Fe – meaning relatively rich in Mg and Fe) types of magma, but in other magmas which are sialic (from Si, Al – relatively rich in Si and Al) in type, the concentration of Mg and Fe may be too low for olivine to form. In the latter case, no crystallization occurs at the olivine energy level, but the magma cools to a lower, pyroxene-forming energy level. It is also possible that some olivine was formed at its appropriate stage but as the temperature falls the olivine may react with the liquid, which is relatively richer in silica, to form pyroxene.

Pyroxene. As has been anticipated, when the temperature of the magma declines and the energy in it decreases to a stage below the olivine level, pyroxene minerals, which commonly contain Ca and Al in addition to Mg, Fe, and SiO_4, are formed. Pyroxenes differ structurally from olivine – which contains isolated complete SiO_4 tetrahedra – in that the silica tetrahedra in pyroxene are linked linearly, one to another, by an oxygen ion that is common to adjacent tetrahedra. In this way a single chain structure of silica tetrahedra arises, see Figure 7. The ratio of O and Si in the chains is 3:1, expressed as SiO_3^{--}. The chains are joined laterally to other chains by ions of Ca, Mg, and Fe.

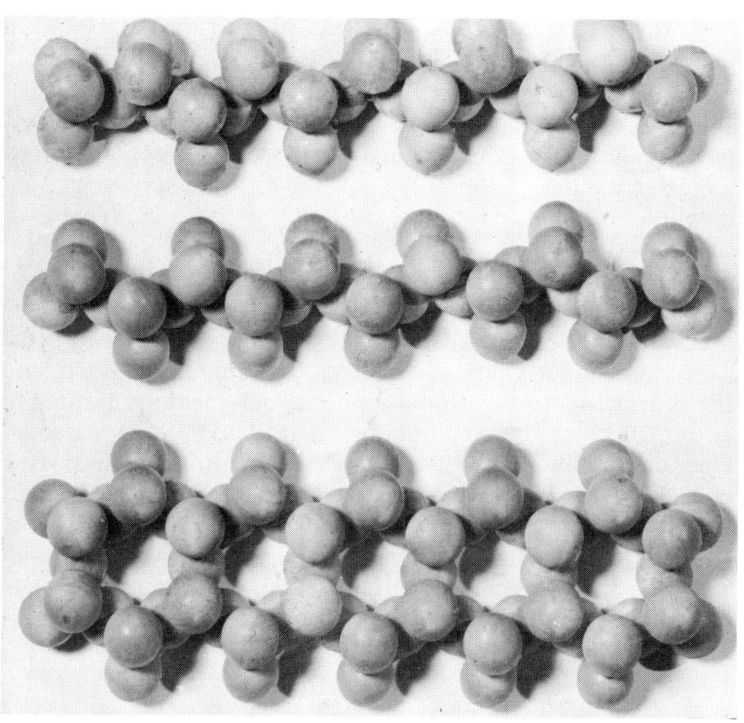

Figure 7. A model of two single silicate chains, the pyroxene type, SiO_3; and one double chain, the amphibole type, Si_4O_{11}. The double chain may be thought of as a product of lateral condensation of two single chains.

Representative formulas for pyroxene minerals are hypersthene, $(Mg,Fe)SiO_3$; diopside, $CaMg(SiO_3)_2$; and augite, $Ca(Mg,Fe,Al)Si_2O_6$. To show the many variations in composition which can occur in pyroxenes requires a long and complex formula; therefore in beginning geology it usually suffices to generalize the formula for the pyroxene minerals as $CaMgFeAlSiO_n$. Cleavage (at 87° and 93°) of pyroxenes yields prism-shaped fragments, the long direction of the cleavage pieces being parallel to the long direction of the chains; the internal structure is thereby reflected in external form and cleavage.

Amphiboles. At a somewhat lower level of temperature and energy than that at which the pyroxenes are formed, a double chain of silica tetrahedra, the Si_4O_{11} ratio develops. See Figure 7. These double chains are linked laterally to each other by Ca, Mg, Fe, and other ions, thereby forming the amphibole or hornblende-type minerals. Representative formulas of the amphiboles are complex, ranging from actinolite, $Ca_2(Mg,Fe)_5(OH)_2Si_8O_{22}$, to common hornblende which is quite lengthy and widely variable, except for the Si_4O_{11} ratio (half that of Si_8O_{22}). For introductory geology students, a formula such as $(OH)CaMgFeAlSiO_n$ is ordinarily sufficiently definitive. Note that amphibole contains OH in addition to the elements commonly present in pyrozene. Cleavage in amphiboles (at 125° and 55°) is clearly long and prismatic, the long direction of the cleavage particles being parallel to the long direction of the chains.

Micas. Descending further in the level of temperature and energy at which crystallization occurs, a sheet structure of linked silica tetrahedra is produced, see Figure 8.

Figure 8. A model of silicate sheet structure, the type in mica and clay, Si_2O_5.

This sheet structure is exemplified by the micas, whose Si: O ratio is 2:5, Si_2O_5, or a multiple thereof. It is developed when 3 of the 4 oxygen ions about each silicon ion in all of the silica tetrahedra (except at broken edges of crystals) act as linking ions to other silica tetrahedra. The formula of muscovite mica may be written $KAl_2(OH)_2Si_3AlO_{10}$, and that of biotite, $K_2(Mg,Fe)_5Al(OH)_4Si_5Al_3O_{20}$, but in introductory geologic work the qualitative formulas suffice, as follows, muscovite, $HKAlSiO_n$, and biotite, $HKMgFeAlSiO_n$. The sheet cleavage of micas is parallel to the sheets of silica tetrahedra. Clay minerals in sedimentary rocks also crystallize in this structure (page 58).

Quartz. The remaining other major type of silicate structure that can be assembled from silica tetrahedra is one in which all 4 oxygen ions of silica tetrahedron act as bridges or links to adjacent silica tetrahedra, Figure 9.

Figure 9. A model of quartz. All four oxygens of each tetrahedron are shared in four adjacent tetrahedra, giving rise to a three-dimensional framework, SiO_2.

This arrangement results in the ratio SiO_2, which is exemplified by quartz. The network is strong in all directions and therefore no good cleavage is developed in quartz.

Feldspars, and ion proxying. One exceedingly important variation, the substitution of Al for part of the Si ions in tetrahedral groups, which can occur in most of the silicate structures will be described in detail for the quartz-type linkage. This modification leads to the formation of feldspar structure. The ion Al is slightly larger than Si, but at the high temperature of a crystallizing magma when all ions possess relatively high kinetic energy and mobility, Al may substitute without structural difficulty for part of the Si ions in tetrahedral structures. Because the size of Al (.57A radius, pages 4,26) is not drastically greater than that of Si (.39 A) the demands of space even at low temperature are not seriously exceeded by substitution of Al for Si.

The electrical balance is upset, however, by substituting Al which carries a charge of 3^+ for Si which carries a charge of 4^+, a deficit of one ($^+$) charge will occur with each substitution of Al for Si. This deficit may be made up by including with the Al substitution another ion which carries a single (+) charge, such as K or Na. Without going into all other intricacies of the substitution, we observe that the common feldspar minerals are developed by this process.

For example, if one out of every 4 Si in 4 tetrahedra is substituted by Al, the change from 4 SiO_2, or Si_4O_8 · is to Si_3AlO_8. By then adding K, or Na, each of which carries a single (+) charge to the group, the electrical balance is maintained, and the minerals orthoclase (and microcline) KSi_3AlO_8, and albite (plagioclase) $NaSi_3AlO_8$, respectively are obtained. These formulas may be written, either as before, or as $KAlSi_3O_8$ and $NaAlSi_3O_8$, if desired, without changing the meaning. The calcium-bearing variety of plagioclase (anorthite) develops by substituting 2 Al for 2 Si and adding one Ca ion which carries 2^+ charges, as follows, $CaSi_2Al_2O_8$, or $CaAl_2Si_2O_8$. Because Na and Ca ions are nearly the same size (Table 1), they may proxy for each other and thereby give rise to a continuous series of Ca-Na compositions in plagioclase feldspar minerals. It is practical, therefore, to generalize the chemical formula of the plagioclases to $NaCaAlSiO_n$, and that for the orthoclase-microcline group to $KAlSiO_n$.

The calcium-rich plagioclase crystallizes from the magma at a higher temperature and energy level than does albite (Na plagioclase). Indeed, calcium-rich plagioclase represents an energy level equivalent to that of olivine and pyroxene.

Although the substitution of Al for Si was illustrated in the quartz-feldspar group of silicates for pedagogical convenience, it should be recognized that this type of substitution occurs also in other silicate groups (aluminous pyroxenes, for one example) and in most minerals. It is a very important principle in crystal and geologic chemistry.

Iron minerals. In addition to the silicate minerals, ordinarily Fe and O ions in magmas combine to form magnetite, the dark magnetic iron ore, Fe_3O_4. Likewise, S ions and Fe ions unite to form the brassy mineral pyrite (fool's gold), FeS_2.

Mineral reaction (Bowen reaction series). Brief comments have repeatedly been made that early-crystallized Mg,Fe silicate minerals characteristically react with the remaining magma. Energywise, these mineral crystals are stable in contact with their parent magma at the temperature of their crystallization. By the withdrawal, however, of such solid, from the liquid, the remaining liquid simultaneously is left richer in silica (SiO_2) than it was originally. Furthermore, as the temperature of the magma continues to fall, the energy relationships (chemical-composition and thermal) that earlier were in balance between solid daughter minerals and parent liquid no longer are in balance.

Consequently, impelled toward reaching another state of equilibrium at the lower temperature, the crystals of MgFe-rich minerals tend to react with the surrounding more silicic liquid. More specifically, as temperature falls, the early formed crystals of olivine $(Mg,Fe)_2{}^{-}SiO_4$, react with the remaining liquid and, by so doing, become a more silicic pyroxene mineral enstatite, $(Mg,Fe)SiO_3$. Such reactions follow the mineral sequence listed on the left side of Figure 4 in a series of discontinuous, stepwise mineral stages in what is called a "discontinuous reaction series".

During the same crystallization interval, Ca-rich plagioclase feldspar also independently crystallizes out at an early stage. Finding itself out of equilibrium with the continually cooling remaining liquid which is becoming increasingly silicic, the Ca-rich plagioclase reacts with the liquid so as to become richer in silica and likewise relatively richer in Na with respect to Ca. That is, a Ca-rich plagioclase feldspar reacts to become ultimately a more silicic plagioclase feldspar richer in Na. Since the plagioclase feldspar compositions vary in an essentially continuous sequence, not by discrete mineral jumps as in the Mg-Fe sequence, the plagioclase reaction is said to be continuous. Both reactions, which go on simultaneously, are named after the late Dr. Bowen who demonstrated their reality.

CRYSTALLIZATION INTERVAL AND ROCK TEXTURE. The time during which the liquid phase of the magma changes completely to solid encompasses the crystallization interval of the magma to rock. The length of crystallization interval, and the fluidity of the magma, are paramount factors influencing the sizes of the mineral crystals, i.e., the texture, of the rock that is formed.

Considering first the liquid phase, the ions comprising it (the liquid) move more freely and more rapidly than do the ions in a solid of the same composition – for example, ions move more freely in water than they do in ice. Likewise, the more fluid the liquid is, the easier will be the movement of ions in it. Relevant also is that the ions in a liquid are distributed relatively at random, or without organization, whereas in crystalline compounds (minerals, for example) the ions-atoms are arranged in a most highly organized, "preferential", relatively fixed combination.

Consequently, as a liquid undergoes complete crystillization the ions must move in the liquid, under relatively weak energy drive, from random distribution to rigorously uncompromising positions in the solid crystal in which adequate space, balanced electric charge, and proper neighbor association, must be satisfied. A long-used analogy to magmatic crystallization is to compare a randomized crowd of individuals attending a football game to ions in a magma. After the game considerable time is required for the individuals to return, moving under relatively weak individual energies within a vigorously jostling crowd, to their stationary domiciles which are rigidly fixed in space.

Similarly, considerable time will be required for the large number of ions necessary to become organized into even a small crystal (a cube-shaped fragment of quartz one millimeter on each edge contains approximately 9×10^{21} ions of silicon, and 18×10^{21} ions of oxygen). To build larger crystals from a solidifying magma, a larger number of ions must move from random distribution to organized position in each crystal. The larger the number of ions required to be organized into each coarse crystal the longer will be the time interval required going from liquid to solid – i.e., a longer interval of crystallization. In other words, a relatively long time (interval) of crystallization is required for relatively coarse crystals (coarse texture) to form.

In addition to the factor of long time, the factor of high fluidity in the magma will promote coarse crystal formation. Mineralizer substances promote such high fluidity.

MINERALIZERS. The emanations from erupting volcanoes, and corresponding products of intrusive igneous rock bodies, show that magmas contain relatively volatile constituents in addition to those which form solid rocks as recorded in the analysis in Table 2. Foremost among the volatile constituents is water, which is steam at the temperature of the magma. Other gases and vapors include sulfur dioxide, SO_2; carbon dioxide, CO_2; carbon monoxide, CO; ammonia, NH_3; ammonium chloride, NH_4Cl; hydrochloric acid, HCl; sulfuric acid, H_2SO_4, and various compounds of hydrochloric, sulfuric, and boric acids. These volatile constituents are called mineralizers because their presence promotes mineral growth as follows: (1) they impart fluidity to the magma, (2) they lower the temperature of its range of liquidity, and (3) they unite with other ions to form additional minerals. The increase in fluidity of the magma favors the easier growth of larger crystals of minerals, shown to a striking degree in the very large crystals of pegmatites. The relatively high concentration of uncommon elements in pegmatite minerals is also believed to be brought about by mineralizer action, in some cases by direct precipitation, but in others by replacement of older, commoner minerals.

Mineralizers build up tremendous pressures at one stage of the cooling and solidification of a magma which often results in violently explosive activity, or in fluid drive which forces fluids long distances through rocks. This source of pressure is explained as follows. At highest temperature of the magma, steam (and other volatiles) are completely and mutally dissolved with the other ions, in the completely liquid magma. The steam does not exist as free steam, but in solution; similar to the gas which is dissolved in carbonated water (soda pop) at low temperature. The steam while in solution in magma liquid exerts no significantly high or extra pressure. The magma, in this condition, does not exhibit extreme pressure.

Suppose that the magma then begins and continues to solidify until a large part of the original liquid becomes solid. At this stage much of the steam which was originally held in solution can no longer be held in solution because the solid rock originating from the cooling magma rejects the steam as it (the rock) becomes solid. The free steam which was rejected is then in contact with very hot solid rock and, following the laws of gas behavior, either expands tremendously in volume or develops terrific pressure.*

PHENOCRYSTS AND PORPHYRIES. Porphyritic texture in igneous rocks may originate in several ways, one of which comes within the scope of this booklet. If a magma contains an exceptionally high concentration of ions with respect to that which produces a normal ratio and sequence of minerals, the magma will probably become saturated first, as it cools, in the ions present in high concentration, with the result that a mineral comprised of them begins to crystallize. More rapid growth, or growth over a longer time interval, of the crystals of the given mineral than of those of the other minerals results in the crystals of the first being larger; therefore they are classified as phenocrysts. Other processes for the origin of phenocrysts are covered in the geology textbooks.

Summary

In summary of the crystallization of a magma, as the magma cools, losing thermal energy, minerals crystallize out, directly or through reaction with pre-existing minerals, as they become insoluble in a sequence of energy levels. A particular mineral represents the availability of adequate concentrations of ions at an appropriate temperature. Therefore, both the composition of a magma and its temperature influence what minerals will result. The common silicate minerals from a magma are olivine, pyroxene, amphibole, mica, and feldspar groups, and quartz.

* The order of magnitude of the pressure may be estimated, but computation of an exact value of the pressure generated is not possible because of the unknown effects of unknown amounts of other substances dissolved in the water (steam), and because properties of steam at the temperatures and pressures of magmas are not well known. However, if water is treated as a mathematically ideal gas, and one volume of liquid water is heated without increase in volume to a temperature at which some magmas are solid, say 750 C. for convenience, a pressure of the order of 10,000,000 pounds per square foot is calculated. This pressure is adequate to balance the weight of a column of average rock more than 10 miles in height. Actual pure water will generate even more pressure than that calculated for the system cited, but the pressure diminishes at a very rapid rate as the volume of the steam increases. The important point to be made here is that steam and gas pressures generated in magmas are entirely adequate to blow away the tops of mountains, and to explode ash and vesiculate pumice in cubic-mile quantity.

Most magmatic minerals are susceptible to reaction with the remaining liquid, which is more silicic than the minerals, to produce increasingly silicic minerals. These reactions may be discontinuous in the series of Mg,Fe-rich minerals, or continuous in the plagioclase feldspars. Magnetite and pyrite are common iron-containing magmatic minerals. Mineralizers play an important role in modifying the texture and mineral composition of the rock.

References for Auxiliary Reading

Principles of Geochemistry, by B. Mason, John Wiley and Sons, New York.

Principles of Chemical Weathering, by W.D. Keller, Lucas Brothers Pub., Columbia, Mo.

Igneous and Metamorphic Petrology, by F.J. Turner and J. Verhoogen, McGraw Hill Book Co., New York, Advanced and technical.

Chemical Weathering

The chemical weathering of rocks is one of the most fascinating topics in geologic science because the weathering process leads to the formation of many earth products which are used by man in his everyday life. Many ore deposits of iron, aluminum, uranium, gold, tin, and the all-important agricultural soil upon which the food of man grows are products of weathering. The extraction by plant roots of nutrients from the soil is a process of soil and rock weathering. Indeed, in a very broad sense, the ingestion and assimilation of food by a human being is a sort of "food weathering".

Unfortunately some trouble is encountered by a few students in beginning geology with the part of the course in weathering, not because the subject of chemical weathering is difficult, but primarily because the most complex part of weathering precedes the simpler parts in the regular sequence of topics in the geology course. This difficulty is avoided in this booklet because we are free to choose the easiest approach to understanding the subject. We will begin with oxidation of rocks and minerals and proceed with simple solution of them in rain water, thence in carbonic acid, and the processes of hydrolysis, hydration, ion exchange and others.

Before discussing the several mechanisms of weathering it is pertinent to inquire why weathering occurs. Fundamentally, weathering processes, which are spontaneous, operate because these changes result, as has been previously stated, in a movement "downhill in chemical energy" of the reactants toward the final products.

Geologically, weathering follows in accord with the so-called law of stability of rocks and minerals. A rock or mineral is stable so long as it remains in the environment (conditions) under which it was formed. A corollary to the law is that when the surrounding environment is changed, the rock or mineral tends to change to a new form which is stable under the new environment.

This may be illustrated by beginning with an intrusive igneous rock. This rock was formed under an environment of relatively high temperature and pressure and a deficiency of free gaseous oxygen and fresh water, and was stable in that environment. When the intrusive rock is then exposed at the earth's surface, either by diastrophic movement or by removal of the rock cover, it finds itself in an environment of low temperature and atmospheric (low) pressure, and in the presence of repeated exposure to, and reaction with, fresh water which contains oxygen and CO_2 gases in solution. Under this new environment, the original silicates, carbonates, and sulfides tend to change into oxides, bicarbonates, and low-energy silicates, and/or to dissolve, insofar as they are soluble.

Oxidation

The most commonly observed reaction in the chemical weathering of rocks is probably that of the oxidation (or rusting) of iron compounds, whereby the red and brown coloration in rocks and soils is produced. The reaction is similar to that undergone by a piece of shiny metallic iron when exposed to moist atmosphere. Oxygen gas dissolves in, and reacts with, water and thence with the iron so that, in effect, each Fe loses three electrons to become Fe^{+++}, and each O gains two electrons to become O^{--}, as follows.

$$4Fe + 3O_2 \rightarrow 2Fe_2O_3 \text{ (hematite)} \quad (1)$$

Hematite, and a hydrated group of iron oxides called limonite (Fe_2O_3-nH_2O), are the red and brown, respectively, minerals in iron rust and red or brown soils and rocks.

The chemically significant change which occurs in oxidation is the loss of electrons, as Fe^{o} to Fe^{+++} , but because oxygen is so commonly involved, the reaction has been called "oxidation". Basically, however, oxygen need not be involved in an oxidation reaction; for example, Fe may be corroded by hot sulfur, S, vapor to be oxidized to Fe^{++} S^{--}, also written FeS. FeS is a black substance differing from pyrite, FeS_2. Note that the iron in the FeS carried a charge of 2^+, the ferrous state, whereas in hematite it carried a charge of 3^+ , a state of higher oxidation, the ferric state. Thus Fe may be oxidized to the ferrous state, Fe^{++}, or to the ferric state, Fe^{+++} . The ferric oxide state is more stable under weathering conditions, hence most common, but ferrous iron is abundant and prevalent in igneous minerals. The generalization that may be made safely here is that iron minerals always tend toward oxidation to Fe_2O_3 and its hydrated compounds when they are exposed at the earth's surface. Multistates of oxidation like those of iron are exhibited by some other, but not all, elements.

OXIDATION PRODUCTS OF PYRITE AND MARCASITE. The oxidation process, combined with usual hydration, can be well illustrated geologically by describing the oxidation of pyrite and marcasite. These iron sulfides are common and widespread in many rocks before weathering. After their weathering, reddish brown spots, stains, and streaks ordinarily remain on the rocks. The reaction, showing only the original materials and the probable final products, may be written as follows.

$$4FeS_2 + 8H_2O + 15O_2 \rightarrow 2Fe_2O_3, \text{ and/or } Fe_2O_3 \cdot nH_2O, + 16H^+ + 8SO_4^{--} \quad (2)$$

The relatively insoluble iron oxide remains on the rock but the soluble sulphuric acid, H_2SO_4, is leached away.

The first reaction of the pyrite (and probably likewise marcasite) as determined from laboratory experiments is apparently oxidation:

$$FeS_2 + 3H_2O \rightarrow Fe(OH)_3 + S_2 + 3H^+ + 3e^- \quad (3)$$

However, on outcrops of iron sulfides undergoing weathering, the mineral melanterite, $FeSO_4.7H_2O$, and acid (sulphuric) waters are observed, suggesting that additional reactions accompany, or soon follow, oxidation as follows.

$$2\,FeS_2 + 16H_2O + 7O_2 \rightarrow 2\,FeSO_4.\,7H_2O + 4H^+ + 2SO_4^{--} \quad (4)$$

Ferric hydroxide, as in (3), can be expected to yield limonite, $Fe_2O_3.nH_2O$, upon rearrangement of the O and H.

Acid water from the oxidizing sulphides reacts with other sulphide to evolve hydrogen sulphide gas, H_2S, which is recognized by its "rotten-egg" odor. H_2S may be partially oxidized to H_2O and yellow, elemental sulphur, S, which, can be observed commonly as crusts coating rock fragments on coal mine dumps, or oxidized completely to H_2SO_4.

Solution by Water

Weathering of rocks by solution in water is illustrated by the removal of beds of rock salt and gypsum which dissolve in rain and ground water that move through the rocks. The process of solution can be clearly demonstrated by stirring a small amount of ordinary table salt, which is a type of processed halite, sodium chloride, NaCl, in a glass of water. The grains of salt, which originally constituted a separate solid phase in the liquid water phase, disappear after contact with the water, and only a single, homogeneous liquid phase, the salt solution remains. The salt is in the liquid solution (it can be tasted) despite the fact that it can not be seen in the liquid, even with a microscope. The solution is characterized by homogeneity in a single phase, which is a criterion we will use to identify other solutions of rock materials.

Gypsum ($CaSO_4.2H_2O$) dissolves more slowly, and in lesser amount, in water than does rock salt. This fact leads us to the realization that solubility is a relative property. Some substances are highly soluble, others dissolve in moderate quantity, and some which hardly dissolve at all are relatively insoluble; those in the last group are popularly said to be insoluble in water, a rather loose but common usage of the term. Quartz sand is an example of a relatively insoluble mineral in water. The quartz sand on the shores of lakes and ocean is washed and stirred countless times by the waves, but so little of the quartz is dissolved by the water that evidence of solubility can not be observed and therefore quartz is said popularly to be insoluble in water.

The concept of solubility and "insolubility" is not difficult, but it is exceedingly important that the student of geology understands and visualizes without second thought what is meant by a substance being soluble or relatively insoluble in a liquid. It is recommended that a mixture of salt and quartz sand be stirred in water to see at firsthand the differences in solubility.

Solution in Carbonic Acid and Carbonation

The dissolving power of water for certain rocks and minerals is increased greatly when carbon dioxide gas, CO_2, is dissolved in the water. Carbon dioxide is liberated when any substance which contains carbon, such as wood, coal, petroleum, natural gas, and most foodstuffs, is burned in air. It is also exhaled by volcanoes. Rain water dissolves a small amount of CO_2 from the air, which contains about .03 of 1 percent CO_2. As the water moves through the ground it picks up more CO_2 in solution because the soil atmosphere is much richer in CO_2, from organic sources, than is the air above ground.

The burning of carbon in air, that is, the combination of carbon with oxygen, is represented by the following reaction:

$$C(\text{solid}) + O_2(\text{gas}) \rightarrow CO_2(\text{gas}) + \text{heat evolved} \qquad (5)$$

The symbolized form used above to describe the reaction by which carbon burns is the type regularly used by chemists. Geology students quickly learn to use the same expressions.

Carbon dioxide gas, regardless of its source, dissolves readily in water; its solubility (amount in solution) is increased as the temperature of the solution is lowered. In contrast to this, the solubility of most solids in water increases with increase in temperature. It is easy to remember from our experience with carbonated soft drinks that CO_2 dissolves more readily at lower than at high temperature. The effervescence ("fizz") in soda pop is CO_2 gas; when the liquid is cold the CO_2 is unnoticed and almost latent in expansive action because the gas is dissolved, but when the carbonated solution is warmed the gas is then less soluble and it evolves from the liquid, frothing and bubbling out of the bottle.

Not only does CO_2 dissolve in water, but these two also react to form carbonic acid, as shown below:

$$H_2O + CO_2 \rightleftharpoons H_2CO_3 \text{(carbonic acid)} \qquad (6)$$

Note that reverse, single-barbed arrows are used to indicate that the reaction can proceed in either direction. When the system is heated the reaction tends to go to the left which releases CO_2 gas, whereas cooling it sends it to the right and tends to form more carbonic acid.

Carbonated water or carbonic acid is a weak acid, and is slightly sour in taste, a common property of acids in general. Acetic acid in vinegar is sour, lactic acid in sour milk is sour in taste, and citric acid imparts the sourness in lemons. More fundamental than in sourness, however, acids are characterized by the fact that they break up, or dissociate, in part, in water to furnish ions of hydrogen,

H^+ ions in water solution*. The H^+ ions are exceedingly tiny (the size of a proton, page 9), but they are aggressively reactive and therefore participate in many chemical and geological changes (reactions).

Now that we understand that carbonic acid furnishes H^+ ions, we may express this fact in symbolic chemical reaction:

$$H_2CO_3 \rightleftharpoons H^+ + HCO_3^- \qquad (7)$$
$$HCO_3^- \rightleftharpoons H^+ + CO_3^=$$

From the foregoing reaction, it is seen that not only are H^+ ions available from H_2CO_3, but also HCO_3^- ions, commonly called bicarbonate ions, and CO_3^{--} ions, called carbonate ions. The HCO_3^- ions are active geologically under more strongly acid conditions, whereas the CO_3^{--} ions are most active under less acid conditions.

The solvent power of carbonic acid is particularly shown by its solution of limestone composed of the mineral calcite, $CaCO_3$, and dolostone (also called dolomite) composed of the mineral dolomite, $CaMg(CO_3)_2$.

$$CaCO_3\text{(calcite)} + H_2CO_3 \rightleftharpoons Ca^{++} + 2HCO_3^- \qquad (8)$$

$$CaMg(CO_3)_2\text{ (dolomite)} + 2H_2CO_3 \rightleftharpoons Ca^{++} + 2HCO_3^- + Mg^{++} + 2HCO_3 \qquad (9)$$

* The symbol "H^+" for hydrogen ion rather than H_3O^+, hydronium ion, is used in this book because H^+ is simpler for non-chemists, is consistent with the explanation of ions, and achieves the same purpose as would the use of hydronium ion. Students who are accustomed to hydronium ion from courses in chemistry will be able to make the adjustment to H^+ ion by subtracting H_2O from H_3O^+.

Acid is commonly used to identify a carbonate, such as calcite, $CaCO_3$. The H^+ ions from acid, such as HCl, hydrochloric acid, react with the CO_3^{--} ions derived from $CaCO_3$ to form H_2CO_3 which, in turn, breaks up into CO_2 gas that bubbles or effervesces visibly in the liquid. The reaction is written:

$$CaCO_3 + 2HCl \text{ (or other acid)} \longrightarrow CO_2\uparrow\text{Gas} + H_2O + CaCl_2 \text{ (or other compound with the other acid)}$$

Dolomite reacts less readily with dilute acid than does calcite, and must be powdered in order to afford effervescence. The two carbonate minerals may therefore be differentiated by applying dilute acid; calcite effervesces in lump form with dilute acid, whereas dolomite in lump resists action but its powder is dissolved.

The foregoing reactions state that when calcite is dissolved in weakly carbonated water, as in rain and soil water, the solution contains Ca^{++} and HCO_3^- ions. When dolomite is dissolved, Mg^{++} ions are likewise present. When such solutions become more concentrated, e.g., by evaporation or as in the ocean, other compounds of Ca and Mg, such as the carbonates of Ca and Mg, are also present.

Water which contains dissolved Ca^{++} and Mg^{++} has properties that characterize "hard water". The presence of dissolved Ca^{++} and Mg^{++} can be demonstrated by the curdling action of that water with soap, and especially by a residue of white "lime", which is composed of carbonates of Ca and Mg, that collects on the inside of the container when the water is evaporated.

The weathering (solvent) effect of carbonic acid water on limestones and dolomites is tremendous. The large underground caves (Carlsbad Caverns, Mammoth Cave, the caves of the Ozarks) and the sinkholes over widespread areas represent dissolved-out portions of limestone. It has been estimated that the land of the entire world is lowered on the average one foot in about 30,000 years by chemical denudation, and the solution of limestone by H_2CO_3 is a large part of this process.

Hydrolysis

Dissolution of minerals may involve a variety of reactions. Silicate minerals, as those in igneous rocks, react with water, i.e., with the H^+ and OH^- ions of water, in a process called hydrolysis (the splitting of water), by which both soluble and relatively insoluble products are formed. The relatively active metal ions, such as Na^+, Ca^{++}, and Mg^{++} go into solution, whereas a complex group of $AlSiO_n$, negatively charged, remains commonly as a framework of scantily soluble clay mineral. Pure water is a powerful and reactive chemical reagent. This statement may be surprising because man intuitively thinks of pure water as a bland, almost inert liquid. Man's body is a watery structure and therefore, water does not react significantly with it, but to feldspar, amphibole, and other silicates, pure water (and aqueous solutions) is probably "public enemy No. 1". Reasons for its reactivity may be explained as follows.

Water, although it is the most common liquid, is actually an unusual chemical reagent having exceptional chemical properties. The large oxygen ion in the H_2O molecule is strongly negative, and the two hydrogen ions are tiny protons (positive) without electrons, therefore they are exceedingly high in electrical charge in relation to their size. Experiments show that the hydrogens occupy positions on the oxygen which are located nearly at two corners of a tetrahedral shape on the oxygen – thus two positive electrical charges are concentrated at two corners, and compensating negative charges are concentrated at the other two corners of what is nearly a tetrahedral pattern on the oxygen.

Neighbor H_2O groups (molecules) take up positions in such a way that the hydrogens of neighbor H_2O molecules are located opposite the negatively charged corners of the first molecule. This pattern is extended indefinitely, although imperfectly, throughout the water, but if the water is frozen, the pattern becomes very well organized. In glacial ice, movement within the ice is not difficult because the tetrahedral shapes of H_2O can move readily past each other at integral distances of the spacing between electrical charges on the hydrogenated oxygens.

Each H_2O molecule (in the liquid) is electrically asymmetrical, having an electrically positive pole on one side of it and an electrically negative pole on the other; hence, water is strongly dipolar and is highly reactive toward both positively-charged and negatively-charged ions. For this reason water should theoretically possess good solvent properties for many substances, and experience confirms that it does dissolve many, many materials.

Furthermore, the highly concentrated positive charge in the hydrogen proton in H_2O can act simultaneously as a bond to other neighbor ions as well as to the oxygen in its own H_2O molecule. This property, the so-called hydrogen bond, enhances greatly the reactivity of water. Water indeed is a powerful and diversely active chemical reagent.

Pure water, even though neutral in terms of acidity, dissociates into H^+ and OH^- ions as follows.

$$H_2O \rightleftharpoons H^+ (10^{-7} \text{ mols per liter}) + OH \; (10^{-7} \text{ mols per liter}) \qquad (10)$$

The preceding reation states that the products of dissociation of water are 10^{-7} mols of both H^+ and OH^- ions per liter of water. A liter of water (amounting to about 1 quart) weighs 1000 grams, and 10^{-7} grams means .0000001 grams. Although this weight of H^+ ions seems, and is, exceedingly small, nevertheless it represents 60,000,000,000, 000,000, active H^+ ions in each liter of water, a number that is large enough to begin an attack on a grain of feldspar or other silicate.

The terms, "acidity, mol, and H^+ ions", have been introduced without definition to maintain continuity in the explanation of hydrolysis. To students who have had no chemistry, it is recommended that they next read the short section on pH (page 67).

It is next appropriate to observe the reaction that occurs when water comes into contact with a silicate mineral. The reaction is readily demonstrated by pulverizing a small piece of nepheline (an igneous mineral, $NaSiAlO_4$), or wollastonite (a metamorphic mineral, $CaSiO_3$), in water and inserting a piece of test paper (Hydrion paper, for example which changes color with change in acidity, Figure 10. When the paper is passed through the slurry of pulverized mineral it darkens immediately, indicating high alkalinity and low acidity. This test shows that a reaction has occurred between the water and the mineral by which the concentration

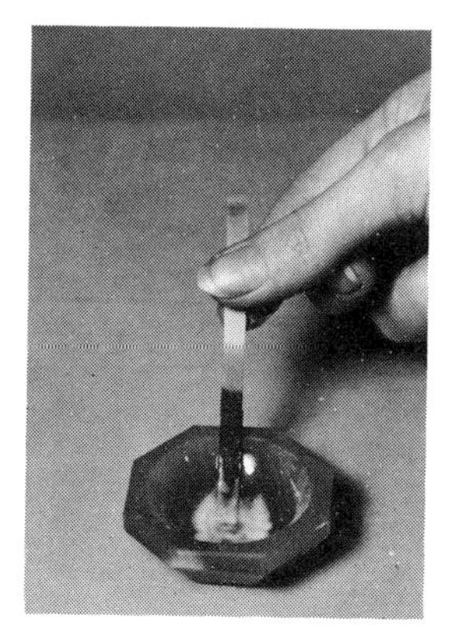

Figure 10. Test paper darkens, showing a decrease in acidity (concentration of H^+ ions) in the slurry of pulverized nepheline.

of H ions from the water has been reduced, and the concentration of OH ions correspondingly increased, in the order of 1000 to 10,000 times. The reaction may be symbolized as follows.

$$NaAlSiO_n \text{ (nepheline)} + H_2O \rightarrow Na^+ + OH^- + HAlSiO_n \quad (11)$$

The preceding equation shows that both the nepheline and water have been broken up or dissociated in the reaction. From the nepheline, Na^+ ions have been set free, but the $AlSiO_n$ group from it has combined with H^+ ions from the water to form a relatively undissociable and insoluble silicate, $HAlSiO_n$, perhaps a clay mineral. The water, prior to the reaction (on the left side of the arrow) was dissociated into equal concentration of H^+ and OH^- ions, but after reaction (on the right), only OH^- ions are shown to be free.*

*The reaction as written is oversimplified. Actually a low concentration of H^+ ions is esent after the reaction but they are dominated by OH^- ions in a ratio of about 10,000,000 H- to 1 H^+.

Although the simple test which has just been described is adequate to demonstrate that the hydrolysis reaction occurs, it does not indicate all of the products of the reaction. Chemical analyses of products from different hydrolysis experiments on silicate minerals show that hydrated silica and hydrated alumina, are also released. Hydrated silica refers to silica combined with variable amounts of water and has been written in several ways: $HSiO_n$, H_4SiO_4, and $Si(OH)_4$, but probably the generalized formula, HSiOn, is adequate to represent it because the exact form in which silica is released during hydrolysis is not well known. Hydrated alumina refers to aluminum oxide and water, written as $Al(OH)_3$, or $Al_2O_3.nH_2O$.

The hydrolysis reaction may be stated in generalized form, with Me to indicate unspecified metal ions, such as K, Na, Ca, in the feldspars, amphiboles, and pyroxenes, and with $HSiO_n$ and $Al(OH)_3$ included, as follows.

$MeSiAlO_n + H_2O \rightarrow Me^+ + OH^-$ $HAlSiO_n$ (usually clay mineral) + $HSiO_n$ + $Al(OH)_3$ (?) (12)

Reaction (12) may be summarized in words of the geologist, as follows: the common aluminum silicates weather (hydrolyze) in water (geologic waters) to produce soluble ions of the metals, soluble silica, relatively insoluble clay minerals, and provisionally, some $Al(OH)_3$. It is observed, furthermore, that the clay minerals tend to remain on the land as soil material, or are washed away as mud, whereas the metal ions and hydrated silica are carried away in solution in stream and ground water and eventually arrive in the ocean which is alkaline in reaction as is expected.

Predicted Responses Of The Common Elements In Generalized Weathering Reactions

By applying to rocks the foregoing responses of chemical elements in weathering, it is possible to predict the products of chemical weathering of rocks in a humid, temperate climate. To facilitate the application, the responses of the elements are condensed into a table (Table 3).

TABLE 3

ELEMENT	REACTION	PRODUCT	SOLUBILITY
Fe	oxidizes	hematite	non-soluble
Fe	oxidizes with hydration	limonite	non-soluble
Fe	occasionally dissolves in carbonic acid	Fe ions	soluble
Ca	dissolves in carbonic acid	Ca ions	soluble
Mg	dissolves in carbonic acid	Mg ions	soluble
Na	hydrolyzes or is carbonated	Na ions	soluble
K	hydrolyzes or is carbonated but is fixed or –	absorbed by clay	non-soluble
	a minor amount escapes as	K ions	soluble
$SiAlO_n$	group hydrolyzes	clay minerals and	non-soluble
		hydrated silica	soluble
SiO_2 as quartz or chert, sl. sol.			scantily soluble

To use the table to predict the weathered products of a rock, the chemical elements present in the rock are derived from its mineral composition, as shown in the following example, a hornblende granite (whose mineral composition ordinarily is orthoclase feldspar, quartz, a small amount of plagioclase feldspar, and hornblende).

Rock	Mineral Composition	Chemical Composition
Hornblende granite	Orthoclase	$KAlSiO_n$
	Plagioclase	$NaCaAlSiO_n$
	Hornblende	$OHCaMgFeAlSiO_n$
	Quartz	SiO_2

The elements composing the minerals are next referred to Table 3, from which the following weathered products of the hornblende granite are ascertained, and grouped according to relative solubilities. (Also see Figure 11 next page)

Relatively insoluble, to soil	Soluble, to streams and ground water
clay minerals	Ca^{++}
quartz	Mg^{++}
limonite	Na^{+}
hematite	$HSiO_n$ (soluble silica)
K sorbed on clay	(Fe^{++}, K^{+}, scarcely dissolved)

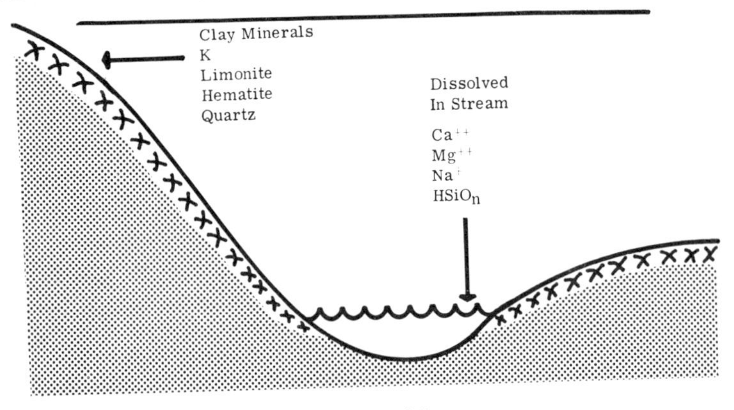

Figure 11

To predict the weathered products of a fine-crystalline rock, a basalt for example, in which the minerals may not be identifiable by the naked eye, its chemical equivalent which possesses a coarse, mineral-identifiable texture is utilized. Thus, the chemically weathered products of a basalt will be the same as those of a gabbro, whose mineral composition can be determined visually, (namely, Ca-rich plagioclase and pyroxene). These mineral formulas, which are known, yield the dominant chemical elements in a basalt responding to chemical weathering.

By using a mineralogical approach, the chemically weathered products of a cherty argillaceous, tan-colored dolomite are predicted as follows.

Rock	Mineral Composition	Chemical Composition
Cherty, tan-colored, argillaceous dolomite	dolomite limonite quartz clay mineral	$CaMg(CO_3)_2$ $Fe_2O_3.nH_2O$ SiO_2 $HAlSiO_n$

Relatively insoluble weathered products	Soluble weathered products
clay minerals quartz limonite	Ca^{++} Mg^{++} $HSiO_n$

ACIDS AND GROWING PLANTS AS AGENTS OF CHEMICAL WEATHERING. The preceding sections have indicated that (1) withdrawal and removal of Ca^{++}, Mg^{++}, Na^{+}, and K^{+}, ions from primary silicate minerals and rocks, and (2) substitution of H^{+} for the metal ions, characterize the chemical weathering of those silicates. It follows logically that active H^{+} ions contributed to the weathering system by acids (see the section on pH) and roots of growing plants will tend to accelerate hydrolysis and other weathering reactions.

Acids. Acids which occur naturally in groundwater include sulphuric, carbonic, and organic varieties such as humic acid. Sulphuric acid commonly originates from oxidizing iron sulphide minerals (as discussed in the section on oxidation). Carbonic acid reaches highest concentrations typically in groundwater because the associated soil atmosphere may be enriched in CO_2 (derived from organic compounds and plant roots) as much as ten times the partial pressure that is present in the air above ground. Humic acid and associated organic acids are commonly derivatives of plant residues.

Growing plants. Living plants, during their processes of growth and metabolism, in which H^{+} ions are released and rock-derived nutrient ions such as Ca^{++}, Mg^{++}, K^{+}, and others, are taken up, become very important agents of chemical weathering. During photosynthesis, when CO_2 and H_2O, plus energy from sunlight, are combined, H^{+} ions are released at the surfaces of rootlets. Such a growing rootlet, acidic and "hungry" for nutrient ions, is shown with a surface atmosphere of H^{+} ions in Figure 11 (see next page). During plant nutrition the root takes up K^{+} or Ca^{++} ions from ambient soil solutions or from exchangable ions weakly held on humus or clay minerals. Such exchange is one H^{+} for one K^{+}, or $2H^{+}$ ion for one Ca^{++}, because the chemical equivalence and balance sheet must be rigorously maintained.

Energy is required to effect a chemical exchange, and the energies available govern what reactions may take

Figure 12 - ENERGY OF ION EXCHANGE

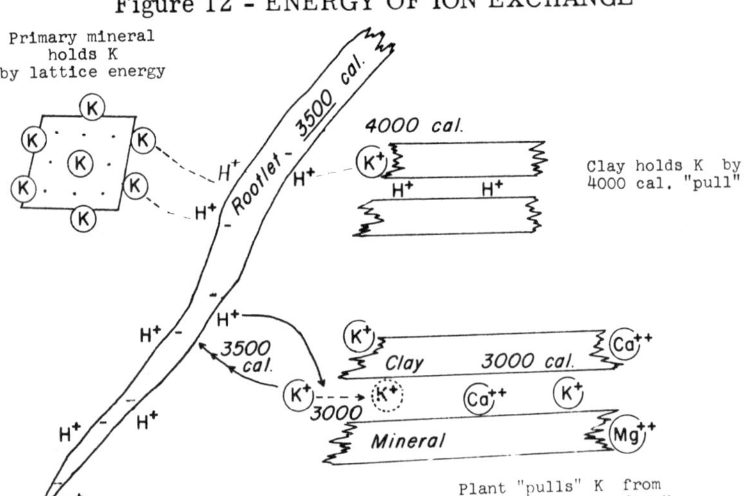

Figure 12. An acid (H^+), "hungry" rootlet is unable to extract nutrient ions that are strongly bonded within fresh primary silicate minerals, or from clay minerals whose bonding energies are higher than that of the rootlet. It does exchange H^+ for nutrient Ca^{++}, Mg^{++}, or K^+ ions from clay minerals that bind the ions less strongly than the bonding energy of the rootlet.

place. Under one set of ion-exchange conditions (which need not be elaborated here) a rootlet may have a mean bonding energy of 3500 calories per chemical equivalent for one K^+, as shown. This rootlet may encounter a tiny particle of K-feldspar (orthoclase) which holds K by very stong crystal lattice bonding energy, far above the exchange energy of the rootlet. The rootlet therefore can not acquire nutrients from fresh feldspar (as in fresh rock) -- only lichen, probably with the aid of symbiotic bacteria, can obtain nutrients from otherwise bare rock.

The rootlet may encounter a particular clay which has a high, mean exchange bonding energy of 4000 calories per equivalent for K^+ --either because it is that particular variety of clay mineral, or because the clay is very scantily stocked with exchangeable K^+ ions. Again, no K^+ will move from clay to rootlet and the plant would not survive in this "infertile" soil.

Alternatively, the rootlet may come in contact with a clay mineral which holds exchangeable ions relatively weakly, e.g., with a mean exchange bonding energy of only 3000 calories per equivalent. Montmorillonite clay, which may be heavily stocked (high ion exchange capacity) with a variety of exchangeable ions, holds its exchangeable ions loosely. Exchange of Ca^{++}, Mg^{++}, and K^+ from the clay for H^+ from the root can therefore take place. By the exchange the plant root has gained nutrients and grows, and the clay has gained H^+, tending to become acid. As such exchange continues, the clay (soil) becomes more acid until exchange can no longer take place because exchange bonding energies come into balance between K^+ and H^+. The clay is no longer "fertile" from the viewpoint of the plant root, and it is acid (acid clay) from a chemical viewpoint.

Such acid clay may be present as a coating on feldspar or hornblende particles of silt or sand sizes. Such acid clay that is coating intimately and continuously a particle of feldspar or hornblende may then react with (i.e., weather) that mineral and extract its K^+ or Ca^{++} in exchange for H^+ from the clay. This reaction and combination of H^+ with $AlSiO_n$ from feldspar or hornblende yields a new "molecule" of clay, a newly weathered product. Such weathering proceeds very slowly, for example, one or more years are necessary to restore naturally developed fertility to soil from tiny rock particles. It is, however, a process of chemical weathering that may occur over wide areas of the earth.

The course of this reaction may be described as follows. Energy from the sun, some 93 million miles away drives a plant (reaction) to supply H^+ to its roots some inches or feet under the earth's surface. These H^+ ions move exchangeably to clay, whereupon the ensuing acid clay weathers originally fresh silicate rocks and minerals to become still more clay. Agriculture, food for man, and an area of botany are the intermediate links within this chain of events categorized as chemical weathering by growing plants.

Effect of Climate on the Types of Clays Produced From the Chemical Weathering of Aluminum Silicate Rocks.

Important different products that result from the weathering of rocks may be correlated with (1) differences in the composition of the parent material (as Table 3 would indicate), and (2) with the type of climate that prevails during weathering. The term "clay mineral" in Table 3 actually refers to a group of clay minerals, among which are the members montmorillonite and kaolinite, whose formulas may be written $HMgAlSiO_n$, and $HAlSiO_n$, respectively. The common clay minerals are discussed in a following section.

EFFECT OF SEMI-ARID CLIMATE. A semi-arid climate receives commonly twenty inches of rainfall, or less, per year, but we shall include also as a semi-arid climate one in which the amount of precipitation is significantly less than what could be evaporated. Under these conditions, when an aluminum silicate rock is wetted, it weathers partly to an AlSiOn complex, and to ions of Ca^{++}, Mg^{++}, Na^{+}, that are dissolved. Under high rainfall these dissolved ions are leached away into the streams, but if drying of the soil and rock intervenes between rains, the dissolved ions remaining in the weathering solution are concentrated as the water evaporates. Under their relatively high concentration these ions will combine with the AlSiOn complex to form montmorillonite clay mineral, which may be written "HMgAlSiOn" with sorbed Ca and Na, as well as K. The Ca, K, and Na ions sorbed on montmorillonite are relatively loosely held. They may be easily exchanged for chemically equivalent amounts of ions, such as H^{+}, and others. Because these easily exchanged ions are relatively abundant on montmorillonite, it (montmorillonite) is said to have a high cation exchange capacity, or alternatively, high base exchange capacity. "Cation" and "base" refer synonomously to metal ions, such as Ca, K, and Na.

Montmorillonite is therefore an excellent source of nutrient ions (Ca, Mg,K), that plant rootlets can pick up in exchange for H ions which are present in the roots as a result of plant metabolism (growth).

Thus, it has been shown how rock weathering in a semi-arid climate produces montmorillonite, a clay mineral rich in Mg, Ca, Na, and K, and favorable in food production. The critical part of this weathering process is the presence of a high concentration of Mg, Ca, and Na ions during drying, and their combination with the AlSiOn complex.

Curiously, but logically, if the weathering solution is not leached away, but the ions are concentrated in a water-logged, warm, weathered zone, montmorillonite is likewise formed.

EFFECT OF A HUMID CLIMATE. In contrast to the previously described environment, if rainfall is abundant and relatively constant, the Mg, Ca, Na, and even much of the K is leached away. Therefore, only H^+ from the water is available to combine with the $AlSiO_n$ complex, thereby forming kaolinite clay minerals, $HAlSiO_n$. Kaolinite possesses a relatively low cation exchange capacity, and is, therefore, pound for pound, not as effective a medium for plant nutrition as is montmorillonite. The climate of particularly the southeastern part of the United States leads to the formation of kaolinite.

EFFECT OF A TROPICAL, RAIN-FOREST CLIMATE. In a warm climate where rain falls practically every day during the entire year, or over a large part of it, and drainage is efficient, leaching of Ca^{++}, Mg^{++}, Na^+, also K^+, and SiO_2 is essentially complete. Note that even SiO_2 is dissolved. Despite its (SiO_2) low solubility, sufficient water moves through the weathering zone that the concentraton of dissolved silica remains below saturation, and at a lower level than will combine with alumina to form clay mineral. Thus, of the originally major constituents of the rocks, only Al and Fe remain as hydrated oxides: $Al_2O_3.nH_2O$, and $Fe_2O_3.nH_2O$. One, or more, minerals of hydrated aluminum oxide, commonly accompanied by impurities, comprise the rock bauxite, which is the common ore of aluminum. It occurs commonly in low-altitude equatorial regions, such as the Guianas of South America. Iron oxide-rich, red clayey rocks, commonly also rich in bauxitic minerals, are called laterites. Hydrated Al and Fe oxides possess low cation exchange capacities and, being scanty in Ca and other ions, are not sustaining producers of proteinaceous foods; they support plants (luxuriant in warm, wet climate) high in wood, fiber, starch, and sugar, such as trees, cotton, and sugar cane. In summary, under the intense leaching of a tropical rain forest climate, all major constituents of rocks except aluminum and iron, which become hydrated oxides, are leached away.

Finding bauxites or laterites in ancient geologic rocks yields a clue to the climate of their origin.

The Clay Minerals

Silicate minerals comprising igneous rocks are characterized by definitive silicate structures which are closely interrelated with the properties of the minerals and the energies of their formation from a magma (pages 21-33). Analogously, the clay minerals, which are the most abundant family of silicate minerals in sedimentary rocks, possess distinguishing structures that are related to the properties and origin of the clay minerals. Clay minerals also may yield information on the climate or depositional environments of their origin, they are a dynamic part of the soil that yields food for man and therefore are of interest to students in geology courses prerequisite to courses in soils and agriculture, and are the source of a large part of earth materials used by man in his technology and culture.

The common clay minerals are silicates of Al, Mg, and Fe built of silica sheets (see page 31). They are micaceous in habit and cleavage, but it is impossible to see their micaceous character with the unaided eye because they are exceedingly fine grained. Most individual clay particles are less than 2 microns in diameter (.002 mm., or somewhat less than 1/10,000 inch) which places them in the colloid size range. Clay minerals are characteristically sticky and capable of being molded, hence plastic, when wetted.

THE KAOLIN GROUP OF CLAY MINERALS. Kaolinite, which is the common representative of the kaolin group, has the generalized formula $HAlSiO_n$, but more specifically is $H_4Al_2Si_2O_9$, or for structural designation, $(OH)_8 Al_4Si_4O_{10}$. From chemical considerations, the important features about kaolinite are that:

(a) it contains H but no K, Na, Ca, or Mg,

(b) the Si:Al ratio is low; 1 Si to 1 Al,

(c) its platy crystals are "sandwiches" or sheets containing 1 layer of octahedral Al and 1 layer of tetrahedral Si ions between O and OH, see Figure 12, and therefore, it is called a 1:1 layer clay mineral,

(d) its crystals do not swell in water, and it possesses relatively low cation exchange properties (explained on pages 5,54, and 57.)

Figure 13 depicts the internal structure of kaolinite crystals two "sandwiches" thick, from two directions of view. Each "sandwich" is 7.13 angstrom units in thickness. Other technical details of the structure are omitted, except to emphasize the 1:1 layer ratio of Si and Al in the kaolinite "sandwich".

Kaolinite is formed by weathering from aluminum silicates where the ratio of concentrations of K^+ ions to H^+ ions is low; this is usually on the acid side. It is derived commonly and readily from the weathering of K and Na feldspars because low concentrations of K and Na ions

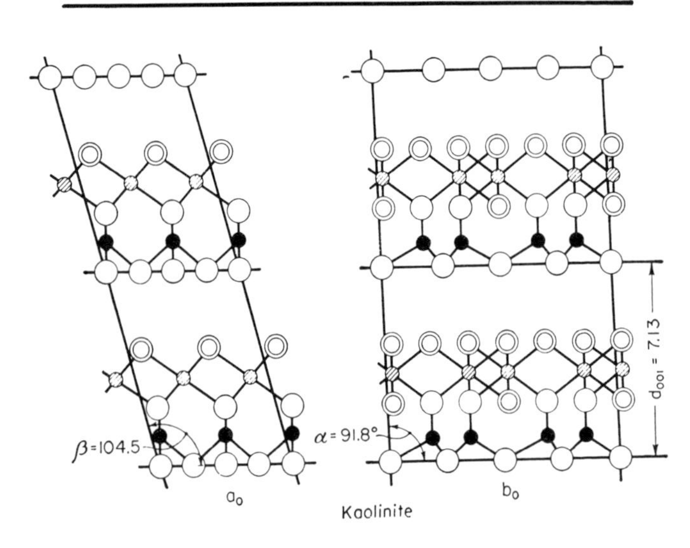

Figure 13. Diagrammatic cross sections through crystals of kaolinite. Large open circles represent oxygen, the large double circles indicate OH, the cross-hatched smaller circles indicate Al, and the solid black ones represent Si. The sheet structure of the clay mineral has been cut in cross section; it extends indefinitely at right angles to the plane of the page. Modified after a drawing by Brindley and others.

stabilize SiO_n in solution, but allow Al to remain insoluble and be relatively enriched. For Ca and Mg silicate parent rocks to produce kaolinite, sufficient leaching is required to maintain always a low concentration of Ca^{++} and Mg^{++} ions which, if they were present, tend to flocculate SiO_2 (cause the SiO_2 to coagulate or gel) and remain. A high content of SiO_2 with Ca and Mg ions tends, on the other hand, toward the development of another structural type, the 2 Si:1 Al layer clay minerals, as follows.

THE SMECTITE GROUP OF CLAY MINERALS. The smectite group of clay minerals possess a 2:1 layer (2 tetrahedral:1 octahedral) structure, and exhibit swelling properties when soaked in water and certain other liquids. The most abundant member of the smectite group is montmorillonite. In it, Mg proxies (see page 32) characteristically in part for Al in the octahedral layer, whereupon the elemental formula for montmorillonite is written HMgAl SiO_n. Commonly also minor substitution of Al for some Si occurs in the tetrahedral layers. The structural formula for montmorillonite may be written: $(OH)_4 (Al,Mg)_4 (Si,Al)_8 O_{20} . nH_2O$, Na, Ca (exch.).

The usual sheet structure for montmorillonite is shown in Figure 14. Properties that characterize montmorillonite include:

(a) it contains Mg, and small amounts of exchangeable Na and/or Ca,

(b) the Si:Al ratio is high: about 2.5 Si to 1 Al,

(c) the "sandwiches" or sheets contain 2 layers of tetrahedral Si to 1 layer of octahedral Al and Mg, Figure 14,

(d) it swells when additional water enters the "intersandwich" space,

(e) it possesses relatively high cation exchange capacity (explained on pages 55,62.)

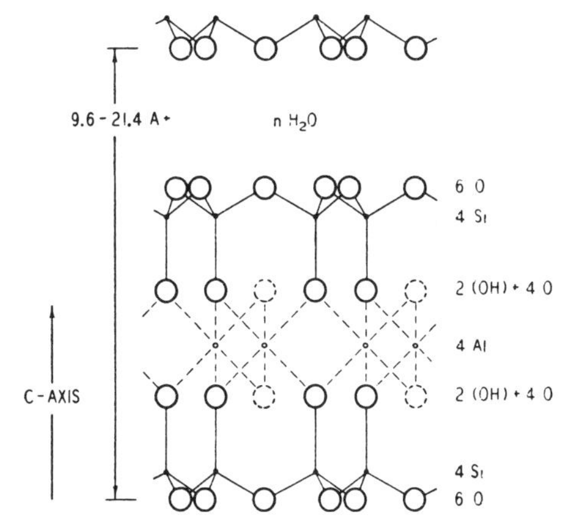

Figure 14. Diagrammatic cross section through a crystal of smectite (montmorillonite). The ions, including those in the easily exchangeable "intersandwich" position, are labeled. Drawing modified after one by Grim, and Hoffman, Endell, and Wilm.

Cation (pronounced "cat"-"ion") exchange capacity is an important property of many substances including clay minerals and zeolites. Cation exchange property in a substance refers to the ability of the substance to permit substitution of part of the cations (ions with a (+) charge, as Na^+ or Ca^{++}) in its structure for other cations. The cation exchange capacity of clay is a measure of the exchange property, expressed usually in chemical equivalents per 100 grams of dry clay. The cations which are most easily exchangeable in montmorillonite are those residing in the "intersandwich" zone, designated in Figure 13 by "nH_2O, Na,Ca". They are loosely held in the structure, and compensate by their electrical charges the deficit of charges arising from the substitution of Mg^{++} (2 plus charges) for Al^{+++} (3 plus charges) within the center, or octahedral layer, layer in the montmorillonite "sandwich". Strongly swelling (when soaked with water) montmorillonite commonly contains Na^+ ions in the intersandwich zone. These Na^+ ions may be replaced by 1 Ca^{++} for each 2 Na^+, or by K^+ for 1 Na^+ ion, etc.

This exchange property is very important in agriculture and plant nutrition because the exchangeable cations on clay minerals (and organic matter) constitute a source of nutrients for plants. The plant rootlets take Ca, K, or Mg from the clay, giving in exchange H ions which are a product of the plant metabolism.

Montmorillonite typically has an exchange capacity of about 90 milli-equivalents per 100 grams of dry clay. One milli-equivalent (m.e.) of Na is in grams .023 (the atomic weight of Na divided by 1000); 1 m.e. in grams of Ca is .020, or 40 divided by 1000 times 2 (because the Ca-- ion has 2 charges); and one m.e. of H is 1/1000 gram. Kaolinite has a cation exchange capacity of 5 to 10 m.e. per 100 grams of dry clay, which is obviously low compared to that of montmorillonite.

Montmorillonite is readily produced from volcanic ash, and also from almost all other aluminous rocks rich in

SiO_2, Ca, Mg, and Fe, which undergo weathering in an environment wherein those ions become concentrated. This concentration of ions may occur in either a dry climate where the water is removed by evaporation, or in a wet region from which there is poor drainage and the soluble ions accumulate to high concentration. Hence, montmorillonite may form near bauxite (an anomalous association) if local drainage varies greatly, or in an arid region which is the antithesis of one in which bauxite is formed.

ILLITE GROUP OF CLAY MINERALS. A group of micaceous clay minerals which is specifically characterized by containing essential potassium is called illite or hydrous mica. Its formula is complex because of variation in composition between different members in the group, but it may be somewhat generalized as $(OH)_4K_y(Al,Fe,Mg)_4(Si_{8-y}Al_y)O_{20}$.

The structure of this group is shown in Figure 15. It has a 2:1 layer structure (2 layers of tetrahedral Si and 1 layer of octrahedral Al) like that of smectite, but illite

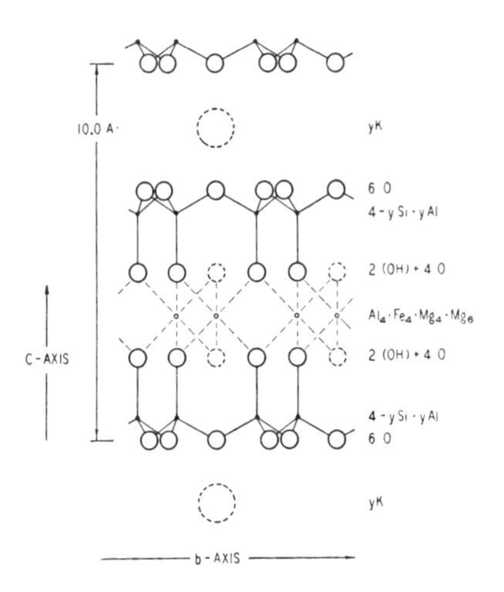

Figure 15. Schematic presentation of the crystal structure of illite. Note the presence of essential potassium (K) in "y" amount. After Grim, Bray, and Bradley.

differs in that it contains significant substitution of Al for Si in the tetrahedral layers, and essential ions of K, in the "intersandwich" positions. The important characteristics of the illite group of clay minerals may be summarized as follows:

(a) contains essential K, also commonly some Mg and Fe,

(b) the Si:Al ratio is about 1.5 Si to 1 Al,

(c) the structure is a 2:1 type,

(d) the clays are non-swelling in water,

(e) the cation exchange capacity is intermediate, between those of the kaolin and montmorillonite group. Potassium is slowly released during weathering.

Illite, which was defined to include a group of micaceous minerals in sediments, may originate in several ways, most commonly by weathering or diagenesis. The weathering of silicate minerals, for example K-feldspar, in an environment where the ratio of concentration of K^+ ions to H^+ ions is comparatively high, and dissolved silica is high, yields illite. These high ratios are significant.

Swelling clays of the 2:1 layer type also may incorporate K^+ within them after deposition (as in the ocean) and reorganize sufficiently to become illite. Such process of illitization of previously swelling clays may be enhanced or accelerated under deep burial by overlying sediments, particularly when temperature of burial rises to about 100° C, and rock pressures are also high.

This process of illitization is in effect, a reversal of one of the weathering processes and accomplishes a change called diagenesis. Ordinarily most processes that change minerals in sedimentary rocks following deposition of the sediment are called diagenetic processes. Although the time factor has long been stressed, it is probably the results of the process that are acutally more basic to diagenesis.

References for Auxiliary Reading

Principles of Chemical Weathering, W.D. Keller, Lucas Brothers Publishers, Columbia, Missouri.

Tropical Soils, E.C.J. Mohr and F.A. Van Baren, Interscience Publishers, New York.

Clay Mineralogy, R.E. Grim, McGraw-Hill Book Co., New York.

pH

Stating that a solution is strongly acid or weakly alkaline, for example, is undesirably vague for describing the degree of its acidity. Instead, the intensity of acidity may be quantified by expressing it by numbers on a simple scale, the pH scale. These numbers, called the pH of the solution indicate the concentrations of the active H^+ ions in the solution.

Concentration means, of course, the amount of a designated substance present in some unit weight or volume, for example, 1 lb. salt per gallon, 120 grams salt per liter, or better still 2 mols (or gram-molecules) of NaCl per liter.

The concept of mol. The concept of mol is essentially an extension of the concept of gram-atom (page 6) to groups of atoms (or atom) comprising molecules, as in a chemical compound. For example, a mol of water, H_2O, is 18 grams of water - the sum of 2 gram-atoms of H(2), plus 1 gram atom of O(16). A gram mol of NaCl (halite) is 58.5 (23 plus 35.5) g of NaCl. Mols are extremely useful forms of measurements relative to chemical combinations because, for example, one mol of H (1g) combines with 1 mol of Cl (35.5 g) to yield one mol of HCL (36.5 g of hydrochloric acid), or 1 mol of Na (23 g) combines with 1 mol of Cl (35.5 g) to yield 1 mol of NaCl (58.5 g).

To return to consideration of concentration in mols, if 58.5 g of NaCl are dissolved in water so that the total volume of solution is 1 liter, the concentration of NaCl is 1 mol per liter of solution. The concentration of a solution is independent of the bulk amount of solution, of course - for example, 3 liters of the preceding solution would contain 175.5 g of NaCl dissolved, or in a teaspoonful of the solution approximately 0.2 g of NaCl will have been dissolved, but each of these solutions will contain the equivalent of 1 mol NaCl per liter, the concentration of NaCl.

Likewise one mol of H^+ ions per liter signifies 1 g of active H^+ ions per liter. One mol of OH^- ions per liter indicates 17 g of OH^- active ions per liter. Now, what is the concentration of H^+ ions in ideally pure water? There is only 1/10,000,000 g of active H^+, and only 17/10,000,000 g of active OH^-! These are otherwise more conveniently stated as 10^{-7} mol of H^+ and 10^{-7} mol OH^-, per liter of pure water. On the other hand, strong acid may contain 0.1 g, or 10^{-7} g, of active H^+ ions per liter, and strong alkali may contain only 10^{-7} g of H^+ ions per liter. Hence, the range in concentration of H^+ ions in geologic solutions is very large, such as 10 followed by 12 zeros, an unwieldly large number. Such large numbers are expressed far more conveniently as powers of 10, in other words, as logarithms of 10 as a base. This scheme of using logarithms is advantageously adopted to expressing concentrations and, in fact, is the basis for the pH scale.

The pH scale. The concentrations of both H^+ and OH^- ions in ideally pure water were reported to be 10^{-7} mols per liter. The acidity, or concentration of H^+ ions, accordingly might have been reported as -7 (the power to which 10 was raised) but the pH scale is even simpler than that. In order to work with simpler positive numbers on the pH scale, rather than more complex negative numbers, the pH scale has been set up arbitrarily, and by definition, as the negative logarithm of the H^+ ion concentration in mols per liter (determined technically against certain standards), see Figure 15. Hence, the pH of ideally pure water is 7 (the negative of a negative-7).

It follows then since the pH numbers are logarithmic (powers of 10), that a solution at pH 4 contains 1000 times (10_3) the concentration of H^+ ions in pure water. At pH 8, the concentration of H^+ ions is 1/100(10^{-2}) of that present in pure water. The pH's of common substances are:

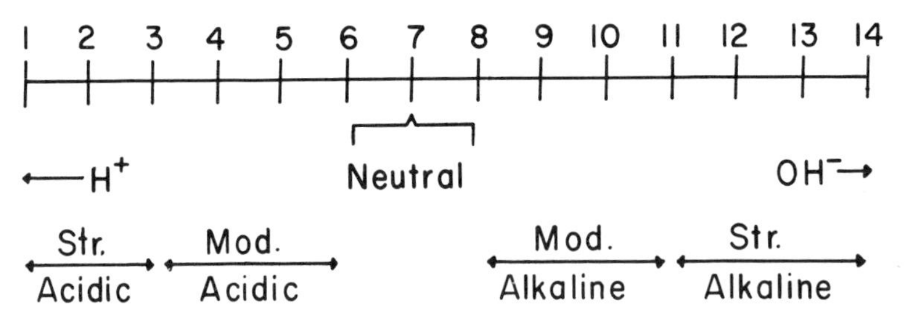

Figure 16

sulphuric acid from oxidizing iron sulphide	1-2
lime juice	about 2
bananas	about 4-6
rain water and common ground water with dissolved CO_2	5 to 6.5
baking soda solution, and ocean water	about 8.2
ammonia water, and alkali-lake water	10-11
strong caustic lye	14

It is easy to remember that pH 7 is neutral, and that the degree of acidity increases toward smaller pH numbers, and alkalinity (increase in OH-, the opposite of acid reaction) increases with larger numbers.

Deposition of the Chemical Sediments

The deposition of sediments which are carried in solution, including those in the colloidal state, occurs by way of processes which are essentially chemical reactions. These reactions are to a considerable degree the reversal of those by which the substances were put into solution. The so-called chemical sediments include limestone, dolostone, gypsum, anhydrite, chert, and other less common rocks and minerals.

Although much limestone is a chemical sediment, many limestones are deposited dominantly by the action of organisms, either directly or indirectly effective.

Deposition of Limestone

The deposition of chemically formed limestone occurs fundamentally by the removal (precipitation) of calcium carbonate, $CaCO_3$, from mobile, soluble calcium bicarbonate, $Ca(HCO_3)_2$, as follows.

$$Ca^{++} + (HCO_3)_2^- \rightleftharpoons CaCO_3 + H_2O + CO_2 \qquad (13)$$

From reaction (13) it is seen that the removal of either H_2O or CO_2 from a solution of $Ca(HCO_3)_2$ results in the deposition of $CaCO_3$. Hence, the causes for deposition of limestone are therefore, in reality, the causes for evaporation of H_2O, or loss of CO_2 gas, regardless of whether the limestone is deposited in the ocean, lakes, caves, or tea kettles. The causes which are most prominent will be reviewed. They are illustrated in Figure 17.

EVAPORATION. Evaporation of water from calcium bicarbonate solutions takes place on a relatively small scale in caves, depositing stalagmites and dripstones, but on a large scale from the surface of the ocean and other bodies of water, resulting in the precipitation of much

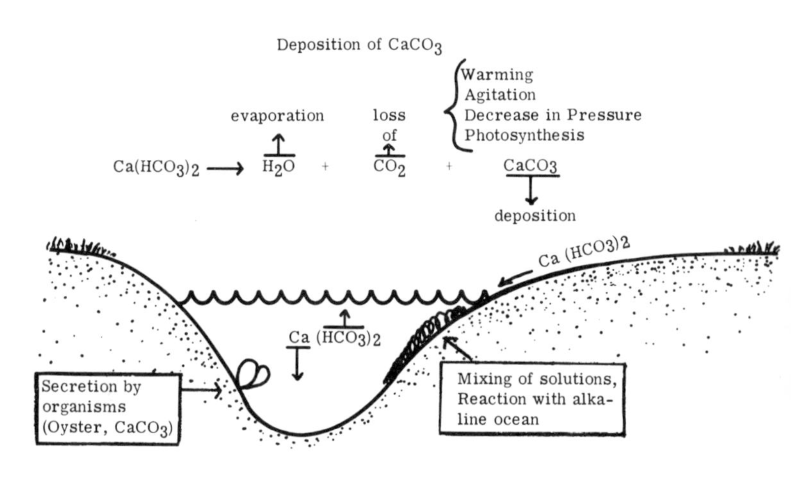

Figure 17

$CaCO_3$ probably as "lime" mud. The amount of water evaporated from the ocean, thence rising to become clouds, and finally falling as rain on the ocean and land is tremendous, and likewise the amount of $CaCO_3$ deposited because of the evaporation of the water is also notably great. Ocean water equivalent to one inch of rain over one square mile can hold in solution (saturated, as it is essentially), before evaporation, about 55 tons of calcite. Hence, a one-inch rain over half the United States, derived from moisture evaporated from the ocean may represent the deposition of a large amount of $CaCO_3$. It is convincing, therefore, that evaporation is an important cause for deposition of $CaCO_3$.

LOSS OF CO_2 FROM Ca-BICARBONATE SOLUTIONS. The CO_2 which is part of calcium bicarbonate solutions may be driven off, thereby reducing the compound to $CaCO_3$ which is relatively less soluble and consequently deposited, as follows.

Heating. Heating a solution of carbonated water decreases the solubility of CO_2 and causes its evolution as a gas – an observation which can be made by watching a carbonated beverage as it is warmed. Heating ocean water will produce a similar result. The ocean is warmed most effectively where sunlight is absorbed as heat by land along the shore or under shallow water in reefs, banks, and gently sloping shelves. Maximum heating of the ocean occurs in equatorial latitudes.

Agitation. Agitation of highly carbonated water, like shaking a bottle of soda pop, releases CO_2. The waves and moving water on lake and ocean shores are adequate evidence that those waters are being agitated. Calcite deposited as cement in sand in the zone of breakers off-shore is evidence of this effect of agitation.

Decrease in pressure. When the gas pressure on carbonated water is decreased, as when a bottle of it under pressure is opened, the CO_2 gas evolves. Accordingly, a decrease in barometric pressure over the ocean, or emergence of water under pressure in the rocks of caves, illustrates mechanisms where the pressure is decreased on $Ca(HCO_3)_2$ solutions in geologic systems.

Photosynthesis of plants. Aquatic plants may take CO_2 out of the water in their metabolism and growth, and thereby cause deposition of $CaCO_3$. This deposition due to extraction of CO_2 is different than secretion of $CaCO_3$ directly.

ALKALINIZING SUBSTANCES. The introduction of alkalinizing substances into a carbonic acid solution tends to neutralize the acid and cause deposition of $CaCO_3$. An example of an alkalinizing substance is ammonia gas, NH_3, which may be given off during the putrefaction of proteinaceous (nitrogen-containing) organisms. The NH_3 reacts with water to produce ammonium hydroxide which is alkaline, see reaction (14).

$$NH_3 + HOH \rightleftharpoons NH_4^+ + OH^- \tag{14}$$

The ocean is alkaline; therefore Ca^{++} brought to it in river water tends to be precipitated as $CaCO_3$.

SECRETION OF $CaCO_3$ BY ORGANISMS. During the growth of certain aquatic animals and plants, $CaCO_3$ may be extracted from the water and built into internal and external anatomical structures, such as bones of vertebrates, shells of oysters, and tests of plant structures. These secretions have accumulated during the geologic past into great thicknesses of rocks.

DEPOSITION BY REPLACEMENT. Various minerals are deposited by processes through which they replace previously existing minerals and even organic substances. Beds and deposits of dolostone (dolomite) have originated by the replacement of calcite, that is, by the introduction and substitution of Mg for half of the Ca ions in a body of calcite. The Ca ions which are replaced move away in solution, and the structure of dolomite, $CaMg(CO_3)_2$, which closely resembles that of calcite, is developed. Dolomite forms also as a direct precipitate from highly concentrated brines.

In addition to replacement occurring by means of substitution of part of the ions in a crystal by others, replacement may occur in more ways, which include the following.

(a) Addition of one or more ions to previously existing minerals, as Cu added to pyrite to form chalcopyrite.

(b) Reaction between the incoming ions and the previously existing mineral by which part of the "host" mineral is incorporated into the replacement, as in the replacement of quartz by micas (vanadium micas, for example).

(c) Complete substitution of one mineral entirely dissimilar to the other substance, as the replacement of silica minerals for wood (cellulose), or the replacement of silica minerals for calcite.

(d) If chemical conditions are reversed, the "host-guest" mineral relationships may be reversed, as when calcite replaces silica minerals, the opposite of the example in (c) preceding.

The basic principle underlying the replacement process is that the guest substance be less soluble within the replacement environment than is the host.

Deposition of Evaporites From Highly Saline Brines

Certain geologic materials are deposited when saturated saline brines are evaporated, or to a lesser extent, when cooled. These materials, commonly called evaporites, include gypsum, anhydrite, rock salt, potassium and magnesium sulfates and chlorides, and sodium sulfate, carbonate, and borate, as well as others. The dissolved substances in the brines may be derived from dissolved products of weathering, igneous emanations, or they may arise from artesian reservoirs, in which case their ultimate origin may be lost in geologic antiquity. Probably the source of the largest amount of evaporites is ocean water which was evaporated in cutoff arms of the ocean where evaporation exceeded inflow of water.

The formation of evaporite deposits is easily demonstrated by dissolving some table salt in water and then evaporating the water; the salt will be recovered in crystals at the bottom and sides of the container. In the same way, beds of rock salt are deposited on the bottom of a lake basin of brine. Cooling of the solution may also decrease the solubilities of the dissolved salts, (the bicarbonates excepted) and initiate their deposition.

Nearly a century ago, an Italian chemist who was interested in evaporites found upon evaporation of water from the Mediterranean Sea that $CaCO_3$ was the main material deposited until nearly four-fifths of the water was gone. At that stage gypsum began to be deposited also, and after about one-tenth the original volume of water remained, solid NaCl began to be formed. From the last residue of remaining liquid, potassium and magnesium sulfates, chlorides, and bromides were obtained.

Colloids in Geology

Colloids and colloidal chemistry are important in geology because clay, iron oxides, and perhaps part of the silica, are commonly transported in, and deposited from, the colloidal state. Because the suspension, transporation, and deposition of colloids are brought about and controlled largely by surface properties of the particles and their reactions, most of which differ from the reactions which are most effective in ionic solution, an additional discussion of the chemistry of geologic colloids is essential.

The colloidal state refers to size of particle, not composition; therefore, it includes particles of any material which range in size roughly between those coarse enough to be seen singly (under the microscope) in suspensions and those which are in so called true or ionic solution – a range in size from about .002 millimeters (2 microns) down to a few Angstrom units in diameter. The surface properties and reactions of particles in this size range are exceedingly important because the total surface area per unit weight increases very rapidly as the particle is reduced in size. For example, the material in a cube as long as the width of an ordinary postage stamp, when pulverized to colloidal-sized cubes 0.1 micron (0.0001 mm.) long has a surface area more than 5,000 square feet, slightly larger than the area of a plot of ground 100 feet by 50 feet. Therefore, since a substance in the colloidal state (a colloid) is mostly surface, it is logical that surface reactions control its suspension and deposition.

We will first inquire into the processes by which colloids are dispersed and suspended, and then reverse those processes which will explain the deposition of the colloids.

The Dispersion and Suspension of Colloids

Clay minerals and grains of silica (quartz), if they occur in sand-size particles or coarser, sink and settle through water. On the other hand, if these minerals, or other rock and mineral particles, occur in the tiny colloid range, it is possible for them to be dispersed (kept separated one from another), and then kept in suspension indefinitely by the so-called Brownian movement, due to the energy (random motion) of the molecules of the water (or other fluid). They are then in stable colloidal suspension. To recapitulate, a colloidal suspension requires particles of tiny colloidal sizes and efficient dispersion of them. If they do not remain dispersed, but clump together, a process called flocculation) they lose the properties of colloids, and the clumps behave like coarse particles. Therefore, the causes and agents of dispersion of colloids are of paramount importance in controlling the behavior of colloids.

The dispersion of clay mineral and silica colloids arises in geologic systems commonly from at least three possible mechanisms.

(1) Colloids may be dispersed when the particles carry strong surface electrical charges of the same kind. Clay minerals and silica particles in water carry net negative electric charges, and because like electric charges repel each other, the clay and silica particles are restrained from touching one another and flocculating – hence, tend to remain dispersed.

The negative charges arise probably from the unsatisfied negative charges from the oxygen ions exposed at the broken surfaces of the mineral particles. These charges attract an atmosphere of positively charged ions and oriented water molecules about each particle, but such details are only corollary to the important concept that negatively charged clay silica colloidal particles repel one another and remain dispersed.

Iron hydroxides are ordinarily positively charged, which likewise establishes mutual repulsion between positive particles and sustains dispersion. It may be anticipated that mixing positively and negatively charged colloids brings about mutual flocculation.

(2) The presence of organic substances, such as aqueous extracts of humus and peat, tend to stabilize the dispersion of colloidal particles. The idea that the organic molecules armor and protect the inorganic colloids from making contact has been proposed as an explanation of the action of the "protective colloids".

(3) A hull of strongly bound hydrated ions and oriented water molecules about colloid particles has been envisaged as another factor in maintaining their dispersion.

Flocculation and Deposition of Colloids

Colloids are flocculated and subsequently deposited if dispersion is no longer maintained. For example, if conditions in the colloidal system change so that the particles no longer repel one another, they collide, remain in contact, and thereby grow by accretion and coalescence until particles larger than colloidal dimensions are formed. These larger particles are no longer held in suspension by Brownian movement but sink through the fluid. This action of clumping together is called flocculation; it may occur quickly, in a matter of a second or less, or it may require days to become significantly effective. The causes for the deposition of colloidal sediments are in reality those effects which permit the dispersed colloidal particles to come into mutual contact and flocculate, as follows.

(1) Decrease of repulsive electric charges. Colloids are flocculated when the mutually repulsive electric charges on the particles are decreased sufficiently that the particles can approach and touch one another. Laboratory experiments have shown that negatively charged clay suspensions are quickly flocculated when fairly high concentrations of any of the common cations, H^+, Ca^{++}, Mg^{++},

Na^+, K^+, are introduced. Accordingly, if colloidal clay is transported from fresh water sources into the ocean or other saline water, flocculation usually begins. It is visualized that not every negative charge on a colloid particle need be neutralized before flocculation takes place, but merely that the electrical fields about the colloid particles are diminished in intensity or modified until the particles can come into contact with one another.

To illustrate the concept that neutralization of the charge on the particles need not be complete, assume for example, two hydrated silica particles moving into close contact. If an external oxygen ion of one of the particles becomes a link or bridge between the silica tetrahedra of the two particles, the two particles may be joined in a quartz-type (page 31) structure without the incorporation or combination with another metal ion from outside. Flocculation by this mechanism differs markedly, therefore, from chemical precipitation and combination between one ion of plus charge with one of minus charge, as in Pb^{++} with S^{--} in galena, or Na^+ with Cl^- in halite. To recapitulate, in the case of colloidal flocculation the effect of an electrical charge opposite in sign to that on the colloid is to diminish the repulsion between particles sufficiently to permit coalescence, whereas in the precipitation of chemical compounds, equivalent (or stoichiometric) amounts of oppositely charged ions are required to complete the precipitation.

Under special circumstances in colloidal systems, reversal of electric charges of the colloid particles, may take place and redisperse them, but these conditions may be disregarded in the geologic systems as discussed.

Mixing of negatively charged colloids with those carrying positive charges causes mutual flocculation.

(2) Removal of protective organic colloid. The removal of protective organic material by oxidation or other means of destruction may permit the then unprotected colloidal

particles to collide, coalesce, flocculate, and build a deposit.

(3) Dehydration of colloidal suspensions. Protective hulls of highly hydrated ions and oriented water molecules about colloidal particles may be destroyed by evaporation of part of the water, or by being captured by a high concentration of ions, as in a salt brine, which is mixed with the original colloid.

Summary

Colloids are flocculated and deposited by removing the agents or causes which keep them dispersed and in suspension.

References for Auxiliary Reading

Principles of Geochemistry, B. Mason, John Wiley and Sons, New York.

An Introduction to Clay Colloid Chemistry, H. van Olphen, Interscience, Pub., New York, 301 pp.

The Colloid Chemistry of the Silicate Minerals, C.E. Marshall, Academic Press, Inc., Publishers, New York.

Chemical Reactions in Metamorphism

Metamorphism of a rock or mineral occurs in accord with the law of stability of rocks and minerals (page 40). When rocks are placed under conditions of high temperature and high pressure, either uniform or directed, the rocks tend to change to new mineral compositions of structures which are stable under the new conditions. The mechanical and thermal energy from the surroundings which is impressed upon the metamorphic system may be taken up in more or less degree by the changing rocks. In this case energy moves "downhill" from the surroundings into the rock being metamorphosed, raising its content of energy with respect to what it was before metamorphism.

If the rocks take up energy of high intensity they respond mineralogically by developing in themselves minerals which have been classified in the category of "high grade" metamorphism – sillimanite and certain garnets, for example. If the energy absorbed is intermediate in intensity, minerals of "middle grade" metamorphism are developed – certain amphiboles, micas, garnets, and staurolite, for example. Low energy and low grade metamorphism are represented typically by chlorite, albite, and calcite, for example. The energy which is stored in the rock during metamorphism may be dissipated later as the rock undergoes weathering.

Although metamorphism is regularly described as change which occurs in solid rocks, as opposed to the liquids present in vulcanism, the rocks during metamorphism are not entirely devoid of fluids. Water, solutions, and aqueous vapors may saturate the pores of the rocks and occupy any larger spaces originally present or that were opened during metamorphism. Igneous emanations provide fluids under pressure to metamorphic systems, and rocks and minerals which contain chemically combined H_2O, as in kaolinite, $H_4Al_2Si_2O_9$, and gypsum, $CaSO_4.2H_2O$, may release steam upon decomposition during metamorphic heating. Carbon dioxide may also be driven off carbonate rocks. Hence, many rocks furnish to themselves mineralizing fluids which provide easy means for the movement of ions through the rock system. In addition to the movement of ions through fluids during metamorphism, evidence indicates that a variable amount of ionic movement or migration occurs through crystalline solids. Materials may be transferred during metamorphism over distances ranging from a fraction of a millimeter to kilometers.

The chemical and mineral changes which rocks undergo during metamorphism are too numerous to mention in entirety, but three of the most important processes which can be used as type examples to which others can be referred will be described.

Recrystallization

One of the important changes during the metamorphism of rocks is the recrystallization of minerals in the rock without significant change in the mineral, or bulk, composition of it – exemplified by the metamorphism of limestone to marble. Ordinary limestone is composed of a more or less homogeneous mixture of crystals of calcite and dolomite, and "impurities", such as clay minerals, tiny carbon particles, iron oxide, and perhaps quartz and pyrite.

If this mineral assemblage comes under pressure at high temperature, energy is added to the ions and compounds, enabling them to move about more readily than before being compressed and heated. Small crystals have higher surface energy than do larger crystals, and therefore give off their ions through the surrounding fluid (liquid or vapor) phase to the large crystals. Originally small crystals in the rock dissolve and vanish; their substance is deposited onto larger crystals which thereby become still larger. However, as crushing and shearing of the rock occurs during metamorphism, the larger crystals may be pulverized into smaller ones, and these may in turn be dissolved and go to other larger crystals. Crystals, or certain portions of them, which are under pressure and strain possess higher energy, and therefore are differentially more soluble, than unstrained portions; the strained parts are dissolved and then deposited on the unstrained portions.

Calcite and dolomite are furthermore particularly well adapted to internal crystalline movement, flowage, and recrystallization because of a highly symmetrical arrangement of ions in the crystal structure, and their perfect rhombohedral cleavage at 120 degrees, or in 3 directions. One part of a calcite crystal may be shifted along a cleavage plane an integral number of ion distances and take up a stable position adjacent to other ions arranged similarly as they were in the first location. This movement may occur easily in all 3 dimensional directions because of the 3 cleavages. The high solubility of calcite and dolomite furthermore conduces to easy healing of fractured surfaces by redeposition after solution. Thus the single mineral species, calcite (or dolomite), in the limestone-marble system may be dissolved and regenerated through many cycles of recrystallization.

When calcite is dissolved, going from smaller to larger crystals, the carbon, clay, and iron oxide impurities are not carried along with the dissolved calcite. Consequently, these impurities tend to be segregated into shapes external to the dissolving and growing calcite crystals. Mechanical movement in the limestone-marble system may stretch, bend, or otherwise distort the segregations of impurities into the bands, blotches, and esthetically beautiful structures preserved in marbles. During recrystallization the "host" calcite mineral rejects from its interior the "guest impurity" which was deposited with it during sedimentation. The calcite thereby is purified to a cleaner, whiter state during metamorphism – recrystallization, and the guest impurities are concentrated into shapes of their own kind. Thus, during metamorphic recrystallization two major changes occur, tendency toward the coarsening of rock texture, and the segregation of different mineral types.

Many other rock-forming minerals in addition to calcite and dolomite undergo more or less recrystallization during metamorphism at temperatures that range from hundreds of degrees centigrade down to surface temperatures as in the recrystallization of rock salt in salt dome flowage structure, to even the freezing point of water, as in glacial ice.

Formation of a New Mineral

Probably the change which most commonly occurs in metamorphism is the formation of a new mineral. Many different examples might be chosen to illustrate this reaction but one of those most directly to the point is the conversion of clay minerals to mica, as in the metamorphism of shale to slate.

It is recalled that shale is a fine-grained sedimentary rock that is generally composed of clay minerals which contain absorbed potassium, quartz, iron oxide, and perhaps calcite or dolomite in small quantities. Shale formations have been traced continuously in the field into metamorphosed zones in which the shale has been changed to slate or mica schist, demonstrating their equivalence. During metamorphism, clay mineral from shale is transformed into mica of slate or schist. More descriptively, the fine-grained, soft clay mineral from shale, which slakes and becomes muddy, sticky, and plastic upon wetting with water is changed into coarser, better crystallized, glistening, crisp, nonplastic, nonabsorbent (to water) mica. Slate repels water so completely that is is used as a roofing material; contrast this property with the slaking of shale. Mica, truly a new mineral, has been formed from clay mineral during metamorphism. Hence the important change during metamorphism of shale to slate is not tighter compaction, but is the change of clay mineral in the shale to mica which characterizes slate. The reaction describing this formation of a new mineral may be written in generalized form as follows.

HAlSiOn (clay mineral) + K (absorbed) + heat —..
..⟶ HKAlSiOn (mica) (15)

If iron oxide or other iron minerals are present in the shale, the Fe, Al, Si and O may combine to form garnet, thereby producing a garnet-mica schist. Feldspar, hornblende, and various other minerals may form from the rearrangement or recombination of ions present in the original rock, or with those introduced by igneous emanations or ionic migration. Tabular and platy shapes of mineral particles, as in gneisses and schists, originate during the growth of the new minerals.

Baking (thermal) Metamorphism

Metamorphism may occur if high temperature is the dominant, or apparently the only agent. Such apparent simplicity is deceptive because usually the effects of static pressure, and either the loss, or introduction, of volatiles also may take place. The metamorphic rock formed in this way is usually a hornfels, which is characterized commonly by being fine grained (even porcellaneous), tightly compact, relatively nonporous, tough, hard, and vitreous in luster. When a search is made for the most fundamental change that occurs during the development of a hornfels and accounts for the physical properties described in the previous sentence, it appears that a reversible change in the structure from crystallinity to glassy condition, and return, augmented by introduction of material, is most basic.

It will be instructive to review the changes that a clay-rich rock undergoes when heated in a dry condition and when permeated by solutions. Montmorillonite, illite, and the extra-hydrated variety of kaolin minerals lose loosely held water without disarrangement of their major crystal structure elements when heated in open air up to about 300 degrees C. At temperatures above about 500 degrees C in the open air the crystalline structure of the kaolin and illite minerals begins to break down into a relatively disarranged, and so-called amorphous, condition in which the clay residue is highly reactive chemically. The chemically reactive residues of the metamorphosed clay may be further rearranged into new minerals as the temperature rises further, or when appropriate mineralizing substances are introduced.

Turning now to natural hornfelses, the thermally metamorphosed equivalents of mudstones, they may show some glassy (amorphous) material in the rock, along with silicification (silica deposited in interstices and replacing other minerals), albitization (albite deposited in and replacing the rock), or sericitization (crystallization of fine-grained muscovitic mica in the rock). It is inferred that those preceding mineral changes can be brought about because thermal metamorphism produces the following effects in rocks.

1. The high temperature of metamorphism increases the energy and mobility of the ions in the rocks, thereby promoting mineral rearrangement and reaction.

2. Destruction of the crystalline structure of original minerals which contain water (either absorbed or chemically combined) and other volatiles, producing an amorphous or glassy structural phase. The non-crystalline portion may remain, or because it is highly reactive chemically, may combine with other components, either residual or introduced, to form a new crystalline phase. During these reactions a tough, closely knit, non-porous, decolorized rock, the hornfels, evolves.

3. Solution and recrystallization of original minerals whose crystallinity are not destroyed by the heat effects.

4. Inducing crystallization in originally amorphous material during the rise in temperature.

Hence, metamorphism that is dominantly thermal produces important changes which are basically structural, that is, changes in the crystalline – amorphous – crystalline states, augmented by mineral reconstitution and replacement.

References for Auxiliary Reading

Principles of Geochemistry, by Mason, John Wiley and Sons, New York, 1952.

The Origin of Metamorphic and Metasomatic Rocks, by H. Ramberg, University of Chicago Press, 1952, Advanced and technical.

Igneous and Metamorphic Petrology, by F.J. Turner and J.Verhoogen, McGraw Hill Book Co., New York, 1951. Advanced and technical.

Deposition of Ore Minerals

Ore minerals of the vein type are believed to be deposited most commonly from migrating liquids and vapors. Rock-forming minerals originate, it is recalled, by crystallizing from a magma, precipitating in surface waters, or evolving during metamorphism. Most metallic ore minerals are chemical compounds which are so scantily soluble in water that the complete chemistry of their deposition involves complexities beyond elementary treatment. Therefore, we shall not treat these topics quantitatively, but we can examine with profit some simple and easily understood chemical reactions and physical chemical changes which illustrate mechanisms by which ore deposition may take place.

Ore minerals are preferentially deposited in limestones and dolomites (carbonate rocks), because the carbonate minerals, calcite and dolomite, are highly reactive toward numerous chemicals, particularly to those which are acid in character, either slightly or strongly. Calcite is highly soluble in acid, primarily because the gaseous (CO_2) part of the reaction escapes from the system, and therefore permits the reaction and dissolution to continue without retardation. This reaction, which will be recognized as the commonly used "effervescence test" for the identification of calcite, may be written as follows.

$$CaCO_3 + 2HCl \text{ (or other anion)} \longrightarrow H_2O + CO_2 + CaCl_2 \text{ (or other anion)} \quad (16)$$

To apply this reaction to the deposition of ore minerals in limestone, let us assume that in the fluids which are to permeate the limestones, the ore element is, in this instance (but not universally so), either (1) a part of the acid substance, or (2) in a compound whose solubility is maintained in the presence of a relatively high concentration of H ions. Then, if and when the ore-bearing fluid comes into contact with the limestone and reacts with it, the acidity of the fluid is neutralized; this reaction reduces the solubility of the ore producing element to such a low value that the ore mineral is deposited (chemically precipitated) in the limestone host rock. The ore mineral grains may be disseminated within the host rock or concentrated in veins.

Even more important are certain other fluids that carry metals in solution as a part of complex ion-groups. Such groups may have low chemical stability, and be sensitive to changes in pH, reaction with wall rock, and change in oxidation potential and temperature. When they encounter a type of wall rock, or environmental conditions with which they can react, the complex ion is divided and the ore metal is deposited.

Ore-bearing fluids may also react with previously deposited metallic or ore minerals. Assume that pyrite has been deposited in the rocks of a given locality and that copper-bearing solutions then permeate the rocks. The copper in solution may react with the pyrite to form chalcopyrite, an ore mineral of copper, as a replacement. One mineral may replace another, the latter being either rock-forming or metal-producing in type, provided that the solubility of the last replacing mineral is less than that of the first one under the conditions of replacement. The conditions of replacement are very important, because mineral A may replace mineral B in one chemical environment whereas in another environment B may replace A, see page 75. Replacement-type ore deposits are very common and very important economically.

Oxidation (or reduction) of ions in solution may decrease the solubility of the elements or compounds sufficiently that they are precipitated and deposited. For example, ferrous iron compounds which are in solution may be oxidized, forming the relatively insoluble red ferric oxide which is then deposited. To the contrary, certain compounds of uranium and vanadium are less soluble in the reduced than in the more oxidized state, and therefore are deposited as ore minerals upon reduction.

Another cause for deposition of ore minerals from solution is a decrease in temperature. Many substances (except for gases, page) are less soluble in aqueous solutions at lower temperature than at higher temperature, and accordingly when hot, ore-bearing solutions rise through cooler rocks near the earth's surface they lose different minerals in a depositional sequence corresponding to their diminishing solubilities. In this way a mineral deposit may be built of overlapping mineral zones which are radially concentric over their source.

If the transportation of the ore compounds depends significantly on the presence of a gaseous component, the diminishing pressure toward the earth's surface conduces toward exsolution and loss of the gas with the result that the ore minerals are deposited.

References for Auxiliary Reading

Economic Mineral Deposits, by A.M. Bateman, John Wiley and Sons, New York.

Solutions, Minerals, and Equilibria, by R.M. Garrels and C.L. Christ,. Harper and Row, New York, 450 pp.

The Rock Cycle

In addition to introducing a few chemical concepts we have considered chemical reactions by which igneous, sedimentary, and metamorphic rocks originate as separate types. Implicit, but not always immediately apparent in these reactions, is that one type or rock may be transformed into another, provided that appropriate energy is imposed upon it. Rock changes may be cyclic if corresponding energies are available, as is illustrated in the accompanying diagram.

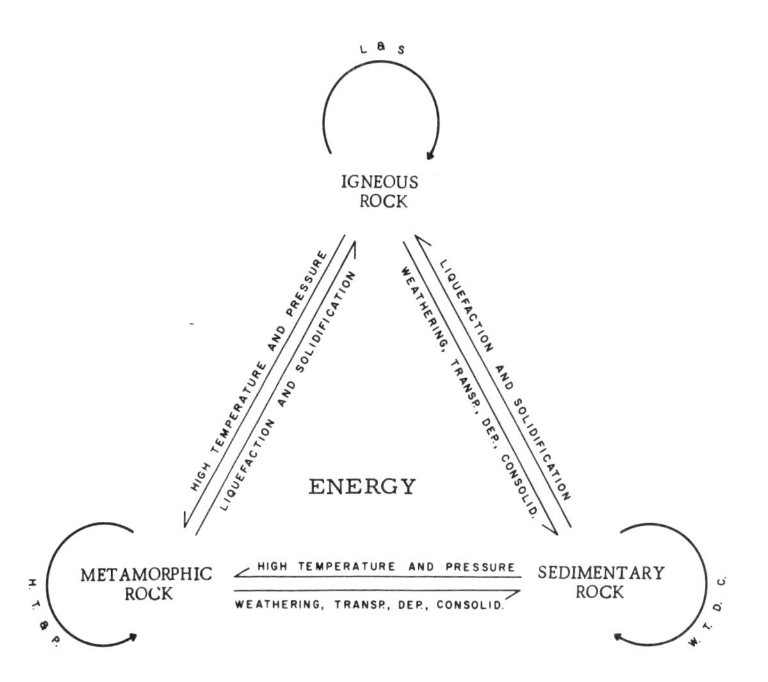

Beginning with igneous rocks (for convenience), when they are weathered, and the products undergo transportation, deposition, and consolidation (abbreviated W.T.D. & C.), sedimentary rocks are formed. Sedimentary rocks may be weathered, etc., to form new sedimentary rocks, or they may be transformed by high temperature and pressure (H.T. & P.) to metamorphic rocks. Metamorphic rocks, in turn, may be further metamorphosed; they may be weathered etc., to become sedimentary rocks; or they may be heated with abundant mineralizers so as to be liquified to a magma and subsequently solidified (L. & S.) to become igneous rock. Indeed, any of the three rock types may be dissolved (liquified) by inclusion in a magma, or by heating in the presence of mineralizers, and lead to a newly formed igneous rock. Igneous rocks may be changed while in essentially a solid state to metamorphic rocks. Thus, each type of rock may be reconstituted to its own type, or be transformed to another type, dependent upon the energy driving the change. Geology never runs out of rocks!

Mineral Age and Environmental History from Isotopes

The ages of certain minerals and rocks may be determined with acceptable accuracy from information on the isotopes of particular elements comprising them. Other isotopic information may yield logical inferences of the temperature, or additional characteristics, of the environment under which the minerals were formed.

In Table 1, page 4, is listed the relative weights of atoms in relation to an arbitrary weight of 12 for carbon, page 6. On page 9, it is stated that most of the mass* of atoms resides in the protons and neutrons of the atomic nucleus. For example, an atom of ordinary hydrogen, contains one proton in its nucleus, and has a weight 1 (more exactly, 1.088). Less abundantly (about 1 part in 6,900 in water) other hydrogen occurs in which one neutron in addition to the proton is present in the nucleus, thus not altering its nuclear charge, but increasing the weight of that atom to 2.00^+; it is called deuterium, ^2H. Even less abundant, about 10^{-7} times the concentration of ^1H in nature, although it may be produced in a reactor, is hydrogen having two neutrons in its nucleus and a weight 3.00+, called tritium, ^3H. Tritium is unstable and spontaneously decomposes, or decays, to ^3He.

By definition, different "varieties" of a single chemical element (as illustrated by hydrogen) which contain the same nuclear charge but differ in atomic weight, are called <u>isotopes</u> of that element. Most chemical elements occur in the earth in several isotopes which may differ considerably

* Weight is the force, which is commonly measured by a balance or scale, that a given mass exerts due to gravitational pull. Hence, the weights of atoms characterize, in common experience, their masses at the earth's surface.

in abundance. Actually, the atomic weight measured for an element may be the average (mean) of the isotopes of that element as they occur in the earth. Isotopes which decay spontaneously (thereby liberating energy) are said to be unstable; those which do not decay spontaneously are stable.

The rate of decay of unstable elements is used to compute the age of the element (and mineral of which it is a part). To illustrate, consider a granite which crystallized in Pre-Cambrian time, possibly more than 1 billion years ago. One of the minerals in the granite, or in pegmatite (very coarsely crystalline) veins associated with it, may be uraninite or pitchblende, an oxide of uranium, which crystallized with the granite. Part of the uranium in the pitchblende is commonly the isotope which has a weight 238, i.e., ^{238}U. This isotope decays spontaneously to one of the isotopes of lead, ^{206}Pb, releasing energy and He.

In the history of the granite and its constituent mineral uraninite, at the time of crystallization of uraninite from solution, ideally, no ^{206}Pb is present. Immediately after the uraninite has crystallized it begins to decay, and simultaneously the quantity of ^{238}U begins to decrease while ^{206}Pb accumulates in increasing quantity. Thus the ratio of ^{238}U to ^{206}Pb present in the mineral is continuously changing, in favor of increasing amounts of ^{206}Pb. The size of the mineral crystals present (or the original amount of uraninite) is not significant, except as it influences the ratio of ^{238}U to ^{206}Pb – the ratio is the significant factor.

The rate of decay of unstable isotopes is exceedingly interesting in the manner which the decay operates. First, the rate is unaffected by chemical reactions, or any other conditions of environment, either natural or artificial, at the earth's surface. Second, and equally significant, the number of atoms that decay during any unit of time is directly related to the number of them that are present at that time. If the number of ^{238}U present originally is large (as in a larger crystal), a large number will decay, designated N, in a given time. However, after half of the original number present have decayed, then only half as many, N/2 will decay thereafter in the same given time. This relationship continues throughout the entire history of decay and thereby permits an age calculation to be made. Note that it was based on the life time of half the atoms present and then on half of those remaining, and so on, continued. For this reason, the time for each half of those present to decay has been called the "half-life" of that isotope.

Half-life reaction may be further illustrated by an example that is "less chemical" than Pb-U isotopes. Suppose that 10,000 coins are shaken in a box for 1 minute, after which the heads facing up are counted. Statistically, 5,000 heads will be up. For the purposes of our example, the 10,000 coins before shaking represented the "U parent atoms" at the time of crystallization of an uraninite mineral. The 5,000 heads represent, after shaking, the "Pb-products of decay" after one half-life, and the 5,000 coins not showing heads represent unchanged U atoms.

Next the 5,000 heads from the first shaking are removed from the box so to simulate separation from uraninite of the Pb atoms; the coins remaining in the box are shaken for one minute (another half life) after which the heads are again counted – this time 2,500 – and removed. There remain 2,500 "parent U" atoms, and 7,500 "daughter decayed Pb" atoms.

For a third time, the 2,500 "parent atoms are shaken for the third half-life of one minute, after which the heads are again counted (1,250). Now 1,250 "parent U" atoms remain, and 8,750 "daughter decayed Pb" atoms have been accumulated. This procedure may be continued as long as U atoms remain.

The significant points of the example are that (1) the number of heads facing up after any one shake depends entirely upon the number of coins present for the "reaction", and (2) each "decay" is in a decreasing half increment, or "half-life".

If the original sample had been 1 million atoms, rather than 10,000, the "decayed" heads would have been proportionately larger (500,000) than the 5,000 as described. Therefore, the original size of the mineral (coin) sample is not significant – it is the ratio between the number of parent atoms and the daughter atoms after any half-life that tells how many half-lives have transpired.

The number of shakings, or half-lives, can easily be calculated if one knows the number of heads (daughter atoms) accumulated, and the number of coins remaining in the shaking box (parent atoms). In minerals, these numbers of atoms are determined by analysis of the minerals.

Clearly then, to calculate the age of a sample of uraninite mineral containing ^{238}U which is decaying to ^{206}Pb, it is necessary to know only the number of daughter ^{206}Pb atoms now present, the number of ^{238}U remaining, and the length of time of one half-life (comparable to the one minute shaking of the coins). These data are substituted in the following equation:

$$\frac{D}{P} = e^{\lambda t} - 1$$

where D is the present number of atoms of the daughter product formed since the time "t" (the age desired), P is the number of atoms of parent isotope in the mineral sample, e is a mathematical constant (number, 2.718), and λ is the rate of decay of the isotope.

The rate of decay, or half lives, of isotopes useful for geologic age dating are long times, as follows.

^{238}U to ^{206}Pb	Approximately	4.5×10^{9}	years
^{87}Rb to ^{87}Sr	"	5×10^{10}	"
^{40}K to ^{40}A	"	1.3×10^{9}	"
^{14}C to ^{14}N	"	5,730	"
^{3}H to ^{3}He	"	12	"

From the foregoing list, it is obvious that U and Rb minerals are used to measure ages of very old rocks, whereas ^{14}C is used on very young geologic deposits (Pleistocene, and others less than 40,000 years old) and archeological materials, and tritium can be used for certain tracer of water-age studies. Although technical details of laboratory manipulation vary considerably between analyses of U-Pb, and Rb-Sr, K-A, ^{14}C, and ^{3}H, the basic application of half-life intervals applies to all of them.

Carbon-14, different from uranium which crystallizes in the solid earth, originates in the far outer atmosphere. It is produced mainly from ^{14}N which, by action of the energy of cosmic radiation, adds neutrons and loses ^{1}H to become ^{14}C. Apparently the ^{14}C soon combines with oxygen to form $^{14}CO_2$ which, at the earth's surface, can enter plants by photosynthesis, or be immobilized in calcite, $CaCO_3$. Whereas a living plant continues in equilibrium with the $^{14}CO_2$ of the atmosphere, upon death its ^{14}C declines in concentration owing to decay of ^{14}C to ^{14}N and loss of electrons. This measured radioactive decay of ^{14}C yields the data on which the time of death of the plant is calculated.

Typical Geological Time Scale
Determined from Isotopes

Cenozoic Era Began			65- 70	million	years	ago
Cretaceous Period Began			130-140	"	"	"
Jurassic	"	"	180-185	"	"	"
Triassic	"	"	225-230	"	"	"
Permian	"	"	280-285	"	"	"
Pennsylvanian	"	"	310-315	"	"	"
Mississippian	"	"	340-350	"	"	"
Devonian	"	"	400-410	"	"	"
Silurian	"	"	420-430	"	"	"
Ordovician	"	"	495-505	"	"	"
Cambrian	"	"	600	"	"	"
Geologic processes began			4,500	"	"	"

Environmental Effects

Variations in environment, such as differences in temperature, salinity, or organic function, may be reflected in the ratios of isotopes present in the associated medium. Such relationships may have been preserved geologically, and inferences of the past be drawn from present-day isotopic analyses. The best elaborated examples are the records of geologically ancient temperatures (paleotemperatures) recorded by oxygen isotopes in $CaCO_3$ of fossils.

The dominant isotope of oxygen at the earth's surface is ^{16}O, but the ^{18}O in marine water, although scanty, is significant as a temperature indicator. In calcium carbonate shells of marine animals, the $CaCO_3$ secreted at warmer temperatures contains relatively more ^{18}O, referred to ^{16}O, than at lower temperatures. Determinations are expressed as deviations from the ^{18}O:^{16}O ratio of standard reference material.

Oysters, for example secrete, as they grow, successive layers of $CaCO_3$ in their shells, somewhat analogous to the annual rings in trees. By analyzing adjacent layers of Ca CO_3 in a shell for oxygen isotopes, information can be obtained on the probable temperature of the water in which the animal lived, and on the extent of temperature changes (seasonal) during the animal's life. The number of summers and winters it lived may be inferred. Inferences may be drawn on ocean temperatures in previous geologic periods, on possible glacial and interglacial stages, and other paleoclimates.

Research on variations in isotopes also of C, S, boron, and the alkaline earths, and possible inferences of them, is currently in progress.

References for Auxiliary Reading

How Old is the Earth? by Patrick M. Hurley, paperback, Anchor-Doubleday and Company, New York, 1959.

Applied Geochronology, by Hamilton and Ahrens, Academic Press, 1965.

Radiocarbon Dating, by W.F. Libby, University of Chicago Press, 1952.

Paleotemperatures of the post-Aptian Cretaceous as determined by the oxygen isotope method, by Lowenstam and Epstein, Journal of Geology, vol. 62, pp. 207-248 (1954).

Measurement of paleotemperatures and temperatures of the Upper Cretaceous of England, Denmark, and the Southeastern United States, by Urey, Lowenstam, Epstein, and McKinney, Bulletin of the Geological Society of America, vol. 62, pp. 399-416 (1951).

SUMMARY

The earth is in effect a combination of a storeroom containing many chemical compounds and minerals, and a laboratory in which a large variety of reactions (geologic changes) are occurring simultaneously in different parts of the laboratory. We have been introduced to the elements, ions, and compounds contained in the storeroom.

Geologic processes such as vulcanism, weathering, deposition of chemical rocks, and ores, and metamorphism are chemical reactions carried out on a huge scale. The reactions in one part of the earth may be the reverse of others in another part. If we knew the conditions that are present at each spot in the earth we might be able to predict what reactions and what products would most likely occur. The fundamental guide in the prediction, that is, to the direction in which the reactions will proceed, is a law of chemistry, never found to have been violated, that a spontaneous reaction proceeds toward a decrease, a "downhill direction," in free energy of the products.

From isotopic analyses of elements in certain minerals,

the ages of minerals may be calculated. The time of crystallization or other genetic mechanisms of key minerals dates the age of geologic events of which the minerals are a part. Isotopic study leads possibly to other inferences of the environment under which geologic materials formed.

An increased knowledge of geochemistry will clarify and advance our understanding of geology.

INDEX